The New Economics of Sustainable Consumption

Energy, Climate and the Environment Series
Series Editor: David Elliott, Professor of Technology, Open
University, UK

Titles include:

David Elliott *(editor)*
NUCLEAR OR NOT?
Does Nuclear Power Have a Place in a Sustainable Future?

David Elliott *(editor)*
SUSTAINABLE ENERGY
Opportunities and Limitation

Horace Herring and Steve Sorrell *(editors)*
ENERGY EFFICIENCY AND SUSTAINABLE CONSUMPTION
The Rebound Effect

Catherine Mitchell
THE POLITICAL ECONOMY OF SUSTAINABLE ENERGY

Gill Seyfang
THE NEW ECONOMICS OF SUSTAINABLE CONSUMPTION
Seeds of Change

Joseph Szarka
WIND POWER IN EUROPE
Politics, Business and Society

Energy, Climate and the Environment
Series Standing Order ISBN 978–0230–00800–7

You can receive future titles in this series as they are published by placing a standing
order. Please contact your bookseller or, in case of difficulty, write to us at the address
below with your name and address, the title of the series and the ISBN quoted above.

Customer Services Department, Macmillan Distribution Ltd, Houndmills, Basingstoke,
Hampshire RG21 6XS, England

The New Economics of Sustainable Consumption

Seeds of Change

Gill Seyfang
University of East Anglia, UK

palgrave
macmillan

First published 2009 by
PALGRAVE MACMILLAN

Palgrave Macmillan in the UK is an imprint of Macmillan Publishers Limited, registered in England, company number 785998, of Houndmills, Basingstoke, Hampshire RG21 6XS.

Palgrave Macmillan in the US is a division of St Martin's Press LLC, 175 Fifth Avenue, New York, NY 10010.

Palgrave Macmillan is the global academic imprint of the above companies and has companies and representatives throughout the world.

Palgrave® and Macmillan® are registered trademarks in the United States, the United Kingdom, Europe and other countries.

ISBN-13: 978-0-230-52533-7   hardback
ISBN-10: 0-230-52533-4   hardback

This book is printed on paper suitable for recycling and made from fully managed and sustained forest sources. Logging, pulping and manufacturing processes are expected to conform to the environmental regulations of the country of origin.

A catalogue record for this book is available from the British Library.

Library of Congress Cataloging-in-Publication Data
Seyfang, Gill, 1969–
    The new economics of sustainable consumption
    seeds of change / Gill Seyfang.
        p. cm. – (Energy, climate, and the environment series)
    "The themes of this book were developed at a conference on Grassroots Innovations for Sustainable Development, held at University College London in 2005" – P.
    Includes bibliographical references and index.
    ISBN 978-0-230-52533-7
        1. Consumption (Economics)–Environment of aspects–Congresses.
    2. Consumer behavior–Congresses. 3. Economics–Sociological aspects–Congresses. 4. Environmental protection–Citizen participation–Congresses.
    5. Sustainable living–Congresses. I. Title.

    HC79.C6S49 2009
    339.4'7–dc22   2008033418

10   9   8   7   6   5   4   3   2   1
18   17   16   15   14   13   12   11   10   09

Printed and bound in Great Britain by
CPI Antony Rowe, Chippenham and Eastbourne

To David

Contents

List of Figures

List of Tables

Acknowledgements

I wish to acknowledge the UK's Economic and Social Research Council support for the research this book is based on, through research grant R000223453 and then through a six-year appointment in the Centre for Social and Economic Research on the Global Environment's Programme on Environmental Decision-Making (M545285002 and RES-545-28-5001). I further acknowledge the Research Councils UK for the award of an Academic Fellowship which has supported the writing of this book.

I am grateful for the time, energy and participation of the individuals and organisations who made this research possible. In particular, I wish to thank Dot Bane at Eostre Organics, David Boyle of the New Economics Foundation, Martin Simon at Time Banks UK, Athena and Bill Steen at the Canelo Project, Mike Reynolds of Earthship Biotecture, Kelly Hart and Paul Koppana in Crestone, Colorado, and Tony Wrench in Pembrokeshire. I am also grateful to the many local food, sustainable housing and complementary currency activists and participants who have shared their experiences for the benefit of this work.

The themes of this book were developed at a conference on Grassroots Innovations for Sustainable Development, held at University College London in 2005, and I wish to thank all the participants and speakers of that day. I co-organised that conference with Adrian Smith from the University of Sussex, and ideas we generated from that event form the basis of a co-authored Chapter 4 in this book and continue to resonate as I develop my work on grassroots innovations; I am grateful for his collaboration. Thanks also to Beth Brocket who provided research assistance during that summer.

Thank you to my friends and colleagues for their interest and enthusiasm, patience and encouragement during the writing of this book, and for feedback and advice throughout the process. In particular, I want to thank my partner David Eastaugh for his limitless support over this period.

Finally, thankyou to David Elliott, series editor, for shepherding my book into his series and offering useful feedback along the way, and to Olivia Middleton at Palgrave Macmillan for help and hand-holding. The publishers and I are grateful for permission to reproduce extracts from the following works:

Seyfang, G. (in press) 'Avoiding Asda? Exploring Consumer Motivations In Local Organic Food Networks', *Local Environment* (reprinted with the permission of the publisher, Taylor & Francis Ltd)

Seyfang, G. and Smith, A. (2007) 'Grassroots Innovations for Sustainable Development: towards a new research and policy agenda', in *Environmental Politics*, Vol 16(4), pp. 584–603 (reprinted with the permission of the publisher, Taylor & Francis Ltd)

Seyfang, G. (2006) 'Sustainable Consumption, the New Economics and Community Currencies: developing new institutions for environmental governance', in *Regional Studies*, Vol 40(7), pp. 781–791 (reprinted with the permission of the publisher, Taylor & Francis Ltd)

Seyfang, G. (2005) 'Shopping for Sustainability: Can sustainable consumption promote ecological citizenship?', *Environmental Politics*, Vol 14(2), pp. 290–306 (reprinted with the permission of the publisher, Taylor & Francis Ltd)

1
Introduction: A Consuming Issue

> Now you can be green and gorgeous, eco-conscious and
> highly fashionable, simply by buying the latest climate-
> friendly consumer products. Never mind marching on
> Whitehall or Downing Street, or giving up flying: all you
> have to do to save the planet is shop (Lynas, 2007: 4)

Shopping to save the planet is big business. The products we buy
and the consumer choices we make are imbued with social and ecological implications, which we are increasingly called upon to consider in a move towards more 'sustainable consumption' patterns.
The burden of managing those impacts rests on the shoulders of
individual citizens, to be weighed up and counted alongside the
many other – perhaps more pressing – concerns of affordability, convenience, availability, fashion, self-expression and taste. In this way,
responsibility for environmental governance and decision-making
in its widest sense is shifting from central government to new sets of
actors and institutions, at a range of scales from international coalitions to individuals (Jasanoff and Martello, 2004; Adger *et al.*, 2003). A
recent consumer book on reducing the greenhouse gas emissions
caused by everyday lifestyle actions, ambitiously claims to be 'the individual's guide to stopping climate change' (Goodall, 2007).

Consumer awareness of environmental issues is slowly rising, but
contradictions remain. A recent study found that while 78% of the
public say they are willing to do more to avert climate change, the
majority were taking only tokenistic actions at present (e.g. recycling)
and were not inclined to question 'sacrosanct' behaviours such as

1

car-driving, flying on holiday, meat consumption and so on (Downing and Ballantyne, 2007). If a 'green consumer' can choose between different models of energy-efficient car, but cannot choose a reliable, accessible, convenient and affordable public transport system, then the scope for individuals to effect societal change is limited from the outset.

Sustainable consumption has been studied from a range of perspectives: economic, sociological, psychological and environmental. This book opens up a new field of enquiry by presenting a 'New Economics' model of sustainable consumption which offers the potential for radical change in socio-economic practices; it challenges many tenets of mainstream policy and individualistic green consumerism. The book examines how an alternative vision of sustainable consumption is practiced through innovative grassroots community action, such as local organic food markets, and community time banks. It investigates how new social institutions and infrastructure are created from the bottom up, to allow people to make more sustainable choices in concert with others. The central aim of this book is to examine some of these 'seeds of change' and assess their potential for growth and influence in wider society, as part of a transition to more sustainable consumption.

Sustainable consumption: a new green agenda

The term 'sustainable consumption' entered the international policy arena in Agenda 21, the action plan for sustainable development adopted by 179 heads of state at the 1992 Rio Earth Summit. This was the first time in international environmental discourse that over-consumption in the developed world was implicated as a direct cause of unsustainability. The proposed solutions included promoting eco-efficiency and using market instruments for shifting consumption patterns, but it was also recommended that governments should develop 'new concepts of wealth and prosperity which allow higher standards of living through changed lifestyles and are less dependent on the Earth's finite resources and more in harmony with the Earth's carrying capacity' (UNCED, 1992: section 4.11). These two proposals – the former suggesting reform and the latter a radical realignment of social and economic institutions – represent competing perspectives of the nature of the problem and its solution, and illustrate some of the tensions inherent in a pluralistic concept like sustainable consumption. Here we will refer to them as 'mainstream' and 'New Economics' perspectives on sustainable consumption (see also Jackson and Michaelis (2003), Jackson (2004b) and Seyfang (2004a) for other reviews of sustainable consumption discourses).

From its auspicious beginnings at Rio, the sustainable consumption agenda has evolved through a range of international policy arenas (see for example OECD, 2002a), and become more widely accepted as a policy goal. The more challenging aspects of its original conception became marginalised as governments instead focused on politically and socially acceptable, and economically rational, tools for changing consumption patterns such as cleaning up production processes and marketing green products. So the policy agenda has narrowed from initial possibilities of redefining prosperity and wealth and radically transforming lifestyles, to a focus on improving resource productivity and marketing 'green' or 'ethical' products such as fairly traded coffee, low-energy light bulbs, more fuel-efficient vehicles, biodegradable washing powder, and so forth. Hence sustainable consumption is implicitly defined as the consumption of more efficiently produced goods, and the 'green' and 'ethical' consumer is the driving force of market transformation, incorporating both social and environmental concerns when making purchasing decisions. As Maniates notes, "'Living lightly on the planet" and "reducing your environmental impact" becomes, paradoxically, a consumer-product growth industry' (2002: 47).

There is widespread agreement that the affluent lifestyles of the developed countries must shift towards more sustainable forms of consumption – although there is not necessarily any consensus about what that might be. Despite a growing consensus at policy level, there is still fierce debate about what precisely sustainable consumption means, among civil society actors and grassroots organisations. A range of different scenarios exist, from exhortations to generate 'cleaner' economic growth, through to the actions of anti-capitalist low-consumption lifestyle activists. In any given sector, wildly different prescriptions for sustainable consumption abound. In housing, for example, sustainable housing might be equally conceived of as high-technology eco-efficient modernity, or alternatively low-impact self-build straw-bale houses that recall a simpler, more self-reliant age (Guy, 1997). Each represents a different idea of what sustainable consumption entails and should achieve, along with equally different prescriptions about what a sustainable society would look like.

In order to comprehend and unravel these contradictions, we need to find a way through the policy debates and conflicting models of sustainable consumption, to find a way of producing simple, coherent and above all, relevant strategies for sustainable consumption. There are a number of important questions to be asked: What drives current consumption patterns? Is it individual tastes and preferences, social institutions and norms, or processes of cultural identification? What links environmental concern with action? How do price and principle compete for consumers' attention when they make shopping decisions? And how can a more radical vision 'New Economics' of sustainable consumption be practised within a mainstream policy landscape?

This book aims to answer these questions by presenting a new synthesis of theory and fresh empirical work which examines sustainable consumption in action. To begin, this introductory chapter briefly sets out the problem and scale of unsustainable consumption, and then reviews current thinking on consumption drivers and the motivating forces which influence consumption decisions. Then two competing models of sustainable consumption are described: a mainstream approach and an alternative, New Economics model, in order to establish the primary theoretical framework for the remainder of the book.

Understanding unsustainable consumption

> Economists see consumption in terms of the generation of utility, anthropologists and sociologists in terms of social meanings, and scientists in terms of the human transformation of materials and energy (Heap and Kent, 2000: 1)

What do we mean by consumption? The answer is not straightforward; it is the completion of economic circuits and the satisfaction of wants; it is the creation and maintenance of identity and lifestyles; it is the using up of resources; and for ecological economists, this resource use is limited by environmental constraints within which all economic and social activity exists. Consumption is, of course, an essential process for all living things; we only achieve a zero-consumption lifestyle when we are dead. So our focus is not on consumption *per se*, but rather on the aspects of it which can be made more socially and ecologically sustainable – by which we mean able to meeting our own needs without compromising the ability of future generations to meet theirs (WCED, 1987).

Global consumption patterns are becoming a topic of increasing concern for politicians, environmentalists and social activists concerned with sustainability. It has become a much-quoted truism that consumption behaviour in developed countries must shift towards a more sustainable form, in order to address the enormous inequalities between rich and poor countries, while respecting environmental limits (UNCED, 1992; WCED, 1987; DETR, 1999). The 1998 Human Development Report describes the gross inequality of consumption patterns across the globe, and notes that while per capita consumption in industrialised countries has risen steadily, at an average of 2.3% annually, over the last 25 years, in Africa, household consumption is actually 25% less than 25 years ago. On a global scale, the 20% of the world's population in the richest industrialised countries accounts for 86% of the world's consumption (measured as private expenditure), while the world's poorest 20% have only 1.3%. The burning of fossil fuels, for example, has multiplied almost five-fold since 1950, and the pollution-absorbing capacities of the environment are threatened. A sixth of the world's land area is now degraded as a result of over-grazing and poor farming practices, and fish stocks are seriously depleted, with almost a billion people in 40 developing countries risking the loss of their primary protein source as a result of over-fishing driven by overseas demand for fish oils and animal feeds (UNDP, 1998).

As climate change has become the most pressing environmental issue facing humanity (IPCC, 2007), so too has the inequity of the consumption patterns which contribute to it been thrown into relief. The risks and benefits of emitting carbon dioxide into the atmosphere are sharply divided among the world's economies, with the developed word contributing the lion's share of emissions, while developing countries face the most dangerous impacts. Carbon dioxide emissions, a by-product from burning fossil fuels, are directly related to consumption levels through the energy used to manufacture, grow, transport, use and dispose of products. While world per capita carbon dioxide equivalent (CO_2e) emissions from fossil fuel use is 4.5t, it varies dramatically across countries, from 20.6 t in the United States, 9.8t in the UK, to 1.8t in Brazil and 0.1t in Ethiopia (UNDP, 2007).

The UK's Climate Change Bill is expected to become law by summer 2008 (DEFRA, 2008), enshrining in national legislation the Kyoto Protocol target of reducing the UK's CO_2 emissions to 60% of their 1990 levels, by 2050 (this goal was first put forward in a 2003 Energy White Paper (DTI, 2003b). This target is intended to stabilise atmospheric concentrations of CO_2 at between 450–550 parts per million, which is assumed to offer a reasonable chance of keeping global warming to below 2°C, so avoiding the worst impacts of rising global temperatures (Schellnhuber *et al.*, 2006). But new scientific evidence is emerging that this target is too low: the 2007/2008 Human Development Report points to the catastrophic impacts climate change will have unless stringent targets of around 80% cuts in greenhouse gas emissions[1] are set and adhered to in developed countries (UNDP, 2007). This translates directly into calls for radical changes in consumption patterns in industrialised nations. The UK Climate Change Bill focuses on the key contributors to the UK's CO_2 emissions, which for consumers relate to household energy use (fuel for heating as well as electrical power) and personal transport (private vehicle use and aviation).

However, the greenhouse gases embedded in what we as a nation *consume* are far greater than that in what we *produce*: developed countries export their carbon emissions to developing countries where manufacturing and processing occurs (Druckman *et al.*, 2007). The Carbon Trust's calculations of per capita CO_2 emissions are based not on production (the nationally-emitted CO_2 divided by population), but rather on consumption (tracking the emissions of all goods consumed in the UK), categorised according to 'high-level consumer need' (Carbon Trust, 2006: 1). A consumption focus highlights the environmental impact of food and other consumer goods

---

[1]Although scientifically incorrect, carbon dioxide emissions are often referred to in the literature as simply 'carbon emissions'. Furthermore, this measure normally includes a range of other greenhouse gases with different global warming potentials (such as methane, nitrous oxide and hydrofluorocarbons), converted to carbon dioxide equivalents. The correct term is therefore 'CO_2e'. However, the UK Climate Change Bill focuses exclusively on CO_2.

and services produced overseas, which are commonly excluded from these calculations, and in turn suggests a different set of carbon-reduction policies to one focused on household energy use and transport. By counting not only direct energy use, but also indirect (embedded) emissions, this analysis reveals that recreation/leisure, space heating, and food/catering are the three categories of consumer need which contribute the most CO_2 to per capita emissions, suggesting scope for reduction in terms of some quite different areas of lifestyle than government production-focused policy attends to.

A focus on consumption as a route to sustainable development reveals much about inequality and inequity which a more traditional production-focused approach would neglect. It calls into question not merely the commerce, business and industry behaviour that economic development is traditionally concerned with, but rather the lifestyles, habits, aspirations and routines of individual citizens and households – an area of life normally considered outside the sphere of regulatory attention. A consumption angle furthermore opens up hitherto neglected arenas of 'non-consumption' decisions, and 'non-market consumption' (Princen, 2002a). By going straight to the heart of modern lifestyles, a consumption focus demands that we examine our most mundane decisions and routines for their impacts and implications, and that we question the economic, cultural and social basis of 21st century consumer societies.

What motivates consumption?

How is consumption behaviour determined and maintained, and how may it be influenced to change? Fundamental to the task of achieving behaviour change is an understanding of what drives current consumption patterns. Within the context of sustainable consumption scholarship, there have been a number of broad-ranging reviews of theories of consumer behaviour, which attempt to map out the theoretical terrain of consumer motivations, most notably Røpke (1999) and Jackson (2004b), each of which provide an excellent interdisciplinary overview of key theories of consumption and consumption drivers, both in theoretical abstract, and in historically concrete examples, drawing on insights from economics, sociology, anthropology, politics, cultural theory and psychology. A comprehensive review of theories of consumer motivation is beyond the scope of this book, and the multitude of approaches can be classified according to one typology or another, depending on the purpose of the specific analysis to hand. In any case, it is axiomatic that divisions between social theories and approaches to consumption are never clear-cut nor absolute, and that whatever analytical design is imposed on the literature is for the purposes of convenience and illuminating a particular dimension of difference. Inevitably there are grey areas and examples that fall in between one category and another, but it is hoped that the overall benefit of structuring the theories outweighs the costs of inaccuracy and imprecision at times. With these thoughts in mind, for the purposes of this book theories of consumption are divided into three broad categories (shown in Table 1.1). The first is a utilitarian approach to consumption, belonging within traditional neo-classical economics,

Table 1.1   Theoretical approaches to consumer motivation

Type of Approach	Scale of Analysis	Decision-making	Consumption is	Example of Tools for Sustainable Consumption
Utilitarian	Individual	Cognitive information-processing on basis of rational utility-maximisation	The means to increase utility	Green product labelling; tax incentives for greener products
Social and psychological	Individual	Response to social contexts and psychological needs	Marker of social meaning, cultural differentiator, and satisfier of psychological needs	Social marketing to 'sell' greener lifestyles as desirable e.g. through celebrity endorsement
Infrastructures of provision	Society	Constrained by socio-technical infrastructure	Inconspicuous, routinised habit	Local food initiatives which bypass mainstream provisioning routes

which examines the behaviour of rational individuals in markets. The second looks at social-psychological drivers of consumption such as status display, group membership, and cultural norms, similarly at the scale of individual consumers. The third takes a societal perspective, and studies the socio-technical infrastructure and systems of provision which determine inconspicuous consumption behaviour. A fundamental distinction is made between individual and societal (structural) theories of consumption behaviour in order to better identify where responsibility lies for changing behaviour, and where the power of decision-making – and the scope for change – lies in each approach. However, following Giddens' structuration theory (Giddens, 1984), it is fully recognised that individuals are at the same time constrained by, and co-creators of, societal infrastructure, and that social institutions are reproduced through the daily actions of individuals. Each of these approaches is briefly reviewed below, exploring their theoretical and practical implications in terms of theories of behaviour change, as a basis for the subsequent discussion of sustainable consumption strategies.

The utilitarian approach

The conventional microeconomic view of consumption is derived in a rather circular fashion from assumptions about individual behaviour. It is axiomatic in neo-classical economics that individuals are rational utility-maximisers, that is to say they calculate and follow the course of economic action which brings them the most utility (benefit, pleasure or satisfaction) that they can afford. A typical microeconomics textbook states 'we assume that consumers seek to allocate their expenditures among all the goods and services that they might buy so as to gain the greatest possible satisfaction. We say that consumers try to maximise their satisfaction, or their utility.' (Lipsey and Harbury, 1992: 37). Individuals consume goods and services in free markets with perfect competition, and it is presumed that this behaviour reveals inherent preferences, and illustrates utility-maximisation, and so consumption acts as an analogue for human happiness or wellbeing. Questions of how preferences are formed, or how decisions are motivated, are sidestepped in favour of a 'black box' view of consumer preferences, so the theory rests simply on making inferences of value, based on consumer behaviour. In this approach, which underpins neo-liberal economic policy, economic growth is considered a prerequisite for development, as it offers greater consumption opportunities and higher consumption levels – a proxy for human wellbeing – overall (DETR, 1999).

The utilitarian model of consumption assumes that decision-making is a linear cognitive process, that is an internal calculation of all available information to decide the course of action which will deliver the greatest utility. From this perspective, analysts 'seek the basis for consumption within the individual, through the mechanism of the satisfaction of needs ... [which] are produced internal psychological and cognitive processes, leading to choices within a marketplace of possibilities' (Wilk, 2002: 6). Therefore, efforts to promote sustainable consumption based on this model tend to rely on initiatives to correct market failures, and ensure that individuals have greater information to enact their consumer sovereignty. For example, the UK's Sustainable Production and Consumption strategy prioritises greater business efficiency and product innovation, consumer information campaigns and voluntary green labelling schemes. These are all initiatives to improve market functioning and information flows to the consumer thereby 'Encouraging and enabling active and informed individual and corporate consumers who practice more sustainable consumption' (DEFRA, 2003b: 6).

Consumer initiatives designed to promote pro-environmental behaviour based on this model similarly appeal to the rational individual actor with information on the impacts of particular behaviour, such as wasting energy. It is hoped that consideration of facts and figures will lead to 'logical' changes in behaviour, particularly where there are clear financial incentives for making the prescribed changes (again, energy efficiency delivers immediate cost savings). A good example of this approach in practice is the UK's 'Going for Green' awareness-raising campaign dating from 1995, and its successor 'Are You Doing Your Bit?' from the late 1990s. These government initiatives sought to provide information to consumers about environmental issues such as global warming and ozone depletion, pollution and resource use, along with advice on simple measures consumers could take to reduce their environmental impacts. They both took an 'information-deficit' approach to changing behaviour, assuming that people behaved unsustainably because they lacked information, and so aimed to overcome that barrier by delivering (expert) information to the lay public. Characteristically of this type of strategy, they achieved little in the way of behaviour change due to a complex range of factors why people failed to take the prescribed courses of pro-environmental action (Blake, 1999). The two campaigns were 'less than half-hearted, and ill-focused' (Environmental Audit Committee, 2003: 34), but led to much discussion of the newly-coined 'value-action gap' between what people claim to care about, and what they act on (Blake, 1999; Kolmuss and Agyeman, 2002). Burgess *et al.* review the literature on public attitude surveys and find that 'the remarkably rapid increase in public awareness of environmental issues and embracing of pro-environmental attitudes is coupled with virtually no substantive changes in behaviour at all' (Burgess *et al.*, 2003: 271).

Social and psychological approaches

A wide range of studies and disciplines have questioned the mainstream model of economic activity, and have sought to better understand what motivates consumers to act as they do, and how that behaviour can be modified to promote more sustainable consumption patterns. Critiques have emerged from the sociological and psychological literature on the drivers of consumption, which aim to help explain why efforts based on the cognitive (information deficit) and market-based approaches to behaviour change have been so ineffective, even where information and pricing has strongly favoured more sustainable consumption. These analyses aim to understand and overcome the well-known 'value-action gap' which describes the disjuncture between knowledge, pro-environmental values and resultant action (see for example Jackson, 2004b). For instance, in a study of the factors which influence environmental commitment, Jaeger *et al.* (1993) found that technical information about specific environmental issues was a weak predictor of activism, as was demographic factors such as age, gender, occupational status. Instead, socio-cultural processes and shared rules, values and networks – ethical values and cultural solidarities – played a strong role in determining environmental commitment. The lessons drawn from this study are that the traditional assumptions about public ignorance and/or confusion about environmental issues are wrong – behaviour will not change simply through the provision of better quality information. This study, and a growing volume of later work from across the social science disciplines, suggests that the core factors which influence consumption decisions have barely been touched by an approach based on flows of expert knowledge to lay consumers – the basis of mainstream sustainable consumption policy (Jackson, 2004b, 2005; Burgess *et al.*, 2003; Røpke, 1999).

Taking as their starting point the social contexts within which consumption takes place, and the psychological needs which consumption is intended to satisfy, these studies conclude that consumption is much more than an economic act and the neo-classical conception of sovereign consumer as rational satisfier of wants is in decline (see, for example, Miller, 1995; Fine, 2002). Beginning with the most economistic of the non-utilitarian approaches to understanding consumption, the body of work known as 'behavioural economics' has shown that individuals do not act like 'rational economic want-satisfiers' in real life, and that this 'bounded rationality' has profound implications for policy (Dawnay and Shah, 2005). It finds, for instance, that in contrast to the principles of neo-classical economics, people's choices are influenced by what other people around us are doing and by social norms, and these can change over time (see Jackson, 2005 for a good review of social-psychological theories of consumer behaviour). Furthermore, norms and routines help to reinforce ingrained (unconscious) habits which 'use little nor no cognitive effort' (Dawnay and Shah, 2005: 5) and so are not subject to the rational cost-benefit calculations which orthodox economics assumes takes place when making consumption decisions. Other insights from psychology and experimental economics reveal that people have intrinsic motivations to want to behave in a public-spirited manner, and value fairness in economic outcomes, but that extrinsic motivations (fines and incentives) can crowd these out, resulting in a loss of value-driven behaviour (Frey and Jegen, 2001). For this reason, the system of donating blood has always been voluntary in Britain, for fear of actually reducing the level of donations by treating it as a commercial transaction rather than a citizenly act. Titmuss (1970) showed that where donors were paid in the US, donations fell and for obvious socio-economic reasons, donors were in poorer health than previously, resulting in lower-quality blood supplies. Another factor influencing people's behaviour is their own expectations of themselves, and a discontinuity between our attitudes and our actions (termed 'cognitive dissonance' by Festinger (1957)) can lead to a revision of the beliefs rather than the behaviour, thereby reversing the conventional assumption that actions follow values. However, making *public* pledges tends to encourage a modification of the behaviour to fit the attitude.

Experimental economics has further revealed that people are loss-averse, and that their 'willingness to accept' compensation for losing an asset far exceeds their 'willingness to pay' to keep it. This lack of parity between gains and losses contradicts neo-classical theory, and results in massive discrepancies between economic valuations of environmental resources, depending on how questions are posed (Pearce and Turner, 1990). Rather than being indicative of irrationality or lack of understanding, these computational 'anomalies' are in fact signifiers of the complex social contexts within which choices are made. Similarly, much has been made of how the framing of a problem influences how people respond (for example an intervention which keeps 80% of people alive is seen as preferable to one which kills 20%) and how intuitive judgements influence behaviour – all of which is anathema to the neo-classical economic model (Kahneman *et al.*, 1991). Finally, the phenomenon of too much choice in a marketplace can result in information overload, confusion, inability to make a decision, anxiety about having made the wrong choice, and general demotivation about the efficacy of our decisions – all crucial issues for sustainable consumers (Levett *et al.*, 2003; Schwarz, 2004).

The key message from this literature is that people do not act as isolated individuals, but rather as people-in-society; we do not respond simply to our innate wants and desires, but also and sometimes overwhelmingly to the influences of our peers and fellow citizens, our unconscious habitual routines and to social norms. As individuals our actions are strongly influenced by those around us, highlighting the importance of social networks, peers and institutions in shaping consumption decisions (Burgess *et al.*, 2003; Jackson, 2004b). These studies demonstrate that people think of others' regard, wish to act for the greater good but only if others do the same, and resist the marketisation of some aspects of economic and social activity. In other words, consumption behaviour is strongly influenced by social pressures and calls to consume differently will be mediated through those contexts.

Taking a step further into sociology and anthropology, others have examined the ways that consumption decisions are intricately entwined with meeting social and psychological needs. Patterns of material consumption exercised through the marketplace embody multi-layered meanings above simple provisioning and the goods and services we consume have enormous cultural significance, for example, aspirational consumption, retail therapy, self-expression, a need for belongingness, self-esteem, self-validation, a political statement, an ethical choice, status display, distinction, loyalty to social groups, identity, and so forth (Douglas and Isherwood, 1979; Bordieu, 1984). Consumption cannot be viewed as technically neutral; it is inextricably linked with values and social meaning, and are signifiers of cultural allegiance and social relationships (Jackson, 2007b). From this perspective, preferences are formed, not within individuals or as endowments, but rather between people in a dynamic manner. Consumption is therefore a moral activity, one that supports and strengthens particular forms of social solidarity, and which is symbolic of collective values and interrelationships (Douglas and Isherwood, 1996). Wilk asserts that 'Consumption is a social code and people consume to fit in or stand out' (Wilk, 2002: 7) and that 'people use goods to communicate to others, to express feelings, and to create a culturally ordered environment' (*ibid.*). Conveying status is one such function. For instance, social standing is commonly signified through the display of expensive material possessions, thereby making conspicuous consumption a desirable activity for its social meaning rather than its instrumental value. Hirsch (1977) uses the term 'positional goods' to refer to those items consumed by the elite, and so desired by the rest of society (they signify one's position in society). Once the goods in question are within the reach of wider portions of society, they lose their appeal, and attention – and desire – turns towards a new elite consumer product, thereby fuelling ever-greater consumption. An example of this is international holidaying to sea-and-sun beach resorts, which was until recently the preserve of the wealthy, and seen as a glamorous, exclusive activity. With cheap flights and international weekend breaks within the reach of the vast majority of westerners, these vacations have become commonplace and are even seen as cheap and brash; there is greater status attached to self-improvement activity holidaying and even to nostalgic returns to domestic camping trips – a reversal of the previous generation's values.

Goods have symbolic value, and the consumption of those symbols is an important aspect of who we are and the social world we make for ourselves (Jackson, 2007b). The fundamental point of these analyses is that efforts to reduce consumption for the rational consideration of the environment are doomed to fail because they do not acknowledge the complex motivations to consume which exist within western societies, and the vital social and psychological functions that consumption provides in terms of expressing identity, a sense of belonging, distinction and so on. These deep-rooted motivations must compete with rational appeals towards sustainable consumption, and as they tap into fundamental social and psychological needs, it is unsurprising that they usually triumph. Only by gaining an understanding of the deeper motivations to consume is it possible to envisage ways to begin meeting those needs from other, less materially-intensive goods and services which can equally well deliver the same intangible benefits. Within this framing of consumption, it becomes possible to envisage a strategy to encourage changes in consumption behaviour through shifts in public values, norms and expectations which have knock-on effects on individuals' actions. Indeed, as Burgess *et al.* assert, the importance of supportive social contexts cannot be overestimated: 'an individual cannot be expected to take responsibility for uncertain environmental risks in a captured market. It is asking too much of the consumer to adopt a green lifestyle unless there is a social context which gives green consumerism greater meaning' (2003: 285).

Employing the formidable armoury and experience of the advertising industry, 'social marketing' is the application of tools and techniques normally associated with influencing consumer behaviour for commercial benefit, to the objective of changing public behaviour for a social good – originally around health and family planning (Kotler and Zaltman, 1971). More recent initiatives have focused on pro-environmental behaviour, and in particular on not simply raising awareness, but fostering community-based, everyday behaviour change through altering contextual (interpersonal and situational) conditions, often in subtle and tightly-targeted campaigns aimed at particular demographic or lifestyle segments of the population (McKenzie-Mohr and Smith, 1999; Barr *et al.*, 2006). Its strategic strength lies in tapping into the unconscious motivations for consumption which the advertising industry have so effectively mined for decades, and planting seeds of behaviour change through new associations and the marketing or 'branding' of pro-environmental behaviour as desirable, and hoping that these small changes will lead to 'tipping points' (Gladwell, 2000) and catalyse wider behavioural transformations. For example, older people might be more receptive to recycling and waste reduction campaigns thanks to their experience of 'thrift' and 'make-do-and-mend' from previous generations, whereas younger population groups might be more receptive to emulating celebrities who choose not to fly, or use reusable shopping bags.

Social marketing has become one of the foundational elements of UK government policy for sustainable consumption, through its five-point model of behaviour change. It aims to *encourage, enable, exemplify* and *engage*, thereby aiming to *catalyse* shifts in attitudes and values, and influence the social context of behaviour, set norms associated with the realm of action, generate a sense of collective endeavour, and recruit the population in moving together towards a more sustainable future (HM Government, 2005). Recent UK government work has concentrated more on segmenting the public into groups of consumers who are, variously, able and/or willing to make more or less significant changes to their lifestyles. The aim is to target different pro-environmental behaviour messages to separate groups of consumers, with the objective of achieving small but potentially catalytic changes across society such as wasting less food, avoiding short-haul flights, installing insulation, etc (DEFRA, 2007a).

The infrastructures of provision approach

The discussion above has focused on motivations for behaviour in individuals, both as cognitive information-processing, and within wider social and cultural contexts. In each case the emphasis is largely on conscious and conspicuous consumption decision-making. A further body of work on consumption behaviour moves outward from the individual to examine collective decision-making and the *creation* and *maintenance* of contextual societal institutions, norms and infrastructure which constrains decision-making. In these cases, it is the routine, the habitual and the *inconspicuous* consumption which is studied. This is referred to here as an 'infrastructures of provision' school of thought on consumption, after Southerton *et al.* (2004) and Van Vliet *et al.* (2005), who examined the case of energy and water utilities. They note that 'institutions and infrastructures actively contemporary patterns of demand' (van Vliet *et al.*, 2005: 6) by entering the home and creating co-dependent relationships between supplier and consumer. The approach can also be applied to other systems of provision (for example food supply chains). Systems of provision are vertical commodity chains (comprising production, marketing, distribution, retail and consumption in social and cultural context) which mediate between and link 'a particular pattern of production with a particular pattern of consumption' (Fine and Leopold, 1993: 4), and this perspective highlights the meso-level infrastructure and institutions which individuals both create and are constrained by, as a form of societal 'structuration' (Giddens, 1984; Røpke, 1999; Sanne, 2002). These systems 'lock-in' individuals to particular patterns of consumption, thereby reducing the choices available to them, and at the same time severely limiting the scope of influence of their purchasing decisions, ensuring the reproduction of the infrastructure. For example houses connected to mains water systems are forced to use pure drinking water to flush toilets, and do not have the capacity to capture and recycle their own rainwater, so ensuring continued dependence on mains water provision. Spaargaren (2003) terms this a 'social practices' approach to sustainable consumption because it examines not simply attitudes or actions or structures, but rather bundles of lifestyle practices in different arenas, such as food, clothing, housing, and so on, which exist in between individuals and societal systems of provision. For example, choices about travel are made not merely on an individual basis, but in relation to wider societal decisions (about investment in infrastructure and so on) which determine the systems of provision and available choices. The resultant practices represent an interface between actor and structure.

Echoing this perspective, Sanne (2002) argues that rather than creatively expressing their identity, consumers are locked in to current socio-technical regimes (often determined by business interests), limiting the available choices they may make, and that they are not necessarily willing consumers at all. Similarly, Shove (2003) examines quotidian household practices such as bathing, and reveals how ever-increasing standards of cleanliness in society counteract moves towards greater efficiency in resource use through norms indicating more frequent washing practices. Consumers are effectively trapped within particular consumption patterns and lifestyle practices by the over-arching social structures of market, business, working patterns, urban planning and development (Sanne, 2002; Røpke, 1999). This has implications for locating agency and allocating responsibility: 'in the social practices approach, the responsibility of the individual towards environmental change is analysed in direct relation with social structure' (Spaargaren, 2003: 690). For instance, Levett *et al.* (2003) argue that while the market defines an ever-expanding range of goods and services to choose from, it cannot, by definition, offer choices external to itself. A person might choose one brand of washing-machine over another because of its greater energy-efficiency, but what they cannot easily choose is to purchase collectively and share common laundry facilities among a local group of residents, or to redefine social conventions to reduce the socially-acceptable frequency of clothes-washing. Within the growing body of literature on societal transitions to sustainability, this level of infrastructure is described as the 'socio-technical regime': namely that set of institutions, technologies and structures which set the rules and parameters within which individual actors may exhibit self-determination (Van Vliet *et al.*, 2005).

Given that current systems of provision prevent significant changes in consumption patterns, what can be done to overcome this limitation? Alternative systems of provision, with associated social and economic institutions and infrastructure, require a foundation in alternative values, development goals, motivations and definitions of wealth (Leyshon *et al.*, 2003). Advocates draw out the political economy of, and richer sociological meanings attached to consumption and point to collective institutions as the source of potential change (Maniates, 2002; Fine and Leopold, 1993), but the shift to new systems of provision is neither easy nor straightforward, given that it involves first contradicting and then challenging existing social institutions and socio-technological regimes. For example, efforts to change infrastructures of provision in the utility industries might suggest a shift to microgeneration and domestic energy-production for greater self-reliance. Southerton *et al.* (2004) investigate initiatives such as these and draw some initial conclusions that indicate a range of unanticipated and at times counter-intuitive consequences, for sustainable consumption (see also van Vliet *et al.*, 2005).

Hence in seeking to make the necessary changes to their consumption patterns, ecologically-motivated citizens 'see that their individual consumption choices are environmentally important, but that their control over those choices is constrained, shaped and framed by institutions and political forces that can be remade only through collective citizen action, as opposed to consumer behaviour' (Maniates,

2002: 65–66). By focusing on socio-technical regimes rather than individual decision-making, one can see that 'in consciously exercising our individual, incremental choices, we have sleepwalked into some larger choices and foreclosed others without even realising it. The market can be an "invisible elbow" shoving us into an unwanted corner, rather than Adam Smith's benign "invisible hand"' (Levett et al., 2003: 47).

Perhaps the most fundamental system of provision which sustainable consumption addresses is that of continued economic growth and the capitalist logic of expansion. Efforts to counteract this continued economic expansion and instigate an economy of 'sufficiency' are, by definition, in opposition to the wider socio-technological regime of society. The transition to a reduced-consumption society 'cuts against patterns of thought and expectation that have been cultivated for generations' (Daly and Cobb, 1990: 373). Røpke (1999) identifies a range of economic factors at play at the macro level. These include the inherent pressures of capitalistic competition and commerce which relies on product innovation and diversification, advertising and want-stimulation, and which have expanded the commercial realm into previously private, domestic areas of life. In practice these trends are revealed as an increasing pace of life and product change, inbuilt obsolescence, deregulated credit and financial services to support growing consumption, and labour market institutions which propagate a 'work and spend' culture (translating productivity gains into higher incomes rather than reduced working time). Schor (1998) focuses on this particular aspect of modern society and concludes that a culture of insatiable desire drives the continual pressure to upgrade, improve, replace and recreate the material conditions of our lives, as witnessed through the modern fashion for personal and property 'makeover' shows, and the commercialisation of the domestic sphere. Similarly, Sanne (2002) finds that modern labour institutions are implicated in the reproduction of this 'work and spend' culture, and that individuals find it difficult to step off the treadmill as many societal institutions are geared to support – and reproduce – it, such as the convention of full-time 40-hour working weeks. Sanne concludes that 'Limited advances can be made by changing consumer habits but further progress demands that the political system overcomes the dogma of economic growth or redefines it in terms of individual welfare of a less material-dominated kind' (Sanne, 2002: 286).

Clearly, these manifold factors operate in concert, reinforcing each other and squeezing out alternative opportunities, in a cycle of continuous consumption which exists as a bedrock of modern economies and societies. Indeed, it is this recently adopted culture of consumer*ism* which is the primary obstacle to sustainability, as social, economic and cultural factors contrive to embed materialistic values and a continual desire to consume more to achieve recognition, fulfilment, and worth. Røpke states 'The account of the driving forces behind the willingness to consume tends to be quite overwhelming: growth in consumption seems to be a very well-founded and understandable trend... Consumption makes sense to people, it concerns very important aspects of life' (Røpke, 1999: 416). Therefore efforts to address consumption issues and promote more sustainable behaviour must be equally multifaceted in their approach, taking an holistic and pluralistic approach which recognises the deeply-rooted social and psychological motivations to consume, as well as the technical and economic drivers.

Nevertheless, alternative systems of provision and social institutions which reject the mainstream imperative for economic growth do exist. Local food initiatives aim to establish new food distribution systems bypassing supermarket supply chains; community currencies aim to value and reward the unpaid work in society, incentivising mutual aid rather than competition; low-impact builders seek sustainable models of development which prioritises self-reliance and reduced consumption. They are all seen by their proponents as embodiments of different sets of values, offering a more sustainable infrastructure within which to conduct lives of sufficiency rather than continual expansion of consumption. How these 'seeds of change' emerge and function in opposition to their wider contexts, and how they might grow to spread their influence into the mainstream, is the core focus of this book.

Competing visions of sustainable consumption

Having reviewed the major theoretical strands of consumption behaviour, and the prospects they hold for encouraging more sustainable consumption, this section examines how those theories have been applied through the brief description of two competing models of sustainable consumption (which are discussed in greater depth in the next two chapters). Our analysis adopts a conceptual framework to organise and usefully separate some of the many strands of thought and practice in sustainable consumption, into two main themes. While the dichotomous model presented is a simplification, grouping together many strands of thought that might otherwise not be considered together, the typology serves to fundamentally distinguish between those that favour incremental change (the mainstream policy approach) and wider, fundamental regime change (the New Economics alternative). These positions are briefly described below, and Table 1.2 summarises the contours of this heuristic device.

Table 1.2 Comparing mainstream and New Economics models of sustainable consumption

	Mainstream Sustainable Consumption	New Economics Sustainable Consumption
Objective	Incremental improvements in resource efficiency; continual economic growth through 'consuming differently'	System-wide changes in infrastructures of provision to reduce absolute consumption levels by 'consuming less'
Mechanism	Sustainable consumers send market signals for sustainably-produced goods and services, which drives innovation and improvement	Collective action reshapes socio-technical infrastructures of provision, creating new systems and non-market alternatives where necessary
Consumers	Individual green consumers	Ecological citizens within communities of place, practice and interest
Progress measured by	Traditional measures of economic growth; consumption as a proxy for utility (happiness)	New measures of sustainable wellbeing; consumption not necessarily related to wellbeing
Theories of consumption	Utilitarian Social/psychological	Utilitarian Social/psychological Infrastructures of Provision
Examples	Green and ethical consumerism; corporate greening of global capitalism; social marketing	Local provisioning e.g. farmers' markets; mutual aid e.g. LETS; self-reliance e.g. low-impact development

It also shows that each of the two models of sustainable consumption rely upon different, but overlapping understandings of consumer behaviour.

The mainstream policy approach to sustainable consumption

In 2003, the UK Government announced its strategy for sustainable consumption and production which it defines as 'continuous economic and social progress that respects the limits of the Earth's ecosystems, and meets the needs and aspirations of everyone for a better quality of life, now and for future generations to come' (DEFRA, 2003b: 10). Two years later the UK government's Sustainable Development Strategy had quietly dropped the explicit imperative for economic growth and replaced it with a guiding principle of achieving a 'strong, stable and sustainable economy' and a call to move towards a 'one planet economy' (HM Government, 2005: 16, 43). But in practice, the policies and tools proposed were much the same with an emphasis on decoupling economic growth from environmental degradation, to be achieved through a range of market-based measures, and calling on informed and motivated citizens to use their consumer sovereignty to transform markets by demanding improved environmental and social aspects of production and product design (*ibid.*). Importantly, this consumer behaviour-change aspect of the strategy relies heavily on the cognitive (information-processing) approach to changing behaviour, and only recently has a more sophisticated – but nevertheless individualistic – social marketing perspective been formally adopted (DEFRA, 2007a). This mainstream policy approach to sustainable consumption has been criticised – not least by the government's own Sustainable Development Commission – on the basis of a number of significant factors which critics claim limit the effectiveness and scope of such a strategy (Porritt, 2003). These include market failures, category errors, disenfranchisement and inequity, and at heart, an inability to address the fundamental problem: 'How can consuming more of anything help us save the planet? The point is to consume less – and no one's going to make any money from that' (Lynas, 2007: 5). Critics therefore conclude that the mainstream approach is limited in scope, flawed in design, and unjust in its objectives. (Maniates, 2002; Sanne, 2002; Seyfang, 2004a, 2005; Southerton *et al.*, 2004; Levett *et al.*, 2003; Holdsworth, 2003; Burgess *et al.*, 2003).

An alternative New Economics approach to sustainable consumption

An alternative theoretical approach to environmental governance and sustainable consumption is proposed by a broad body of thought known collectively as the 'New Economics' (Ekins, 1986; Henderson, 1995; Daly and Cobb, 1990; Boyle, 1993). The New Economics is an environmental philosophical and political movement founded on a belief that economics cannot be divorced from its foundations in environmental and social contexts, and that sustainability requires a realigning of development priorities away from the primary goal of economic growth towards wellbeing instead (Jackson, 2004a). It also stresses the benefits of decentralised social and economic organisation and local self-reliance in order to protect local environments and economies from the negative impacts of globalisation (Jacobs, 1984; Schumacher, 1993). Although its traditions go back much further (Lutz, 1999), the UK's New Economics Foundation was founded in 1986 to promote these ideas in research and policy (Ekins, 1986). At the same time, theorists such as Jackson (2007a), Ekins (1986), Max-Neef (1992), Douthwaite (1992), and O'Riordan (2001) are pursuing these ideas within the academic world, for instance by developing new measures of wellbeing, seeking to understand consumer motivations in social context, and debating how an 'alternative' sustainable economy and society might operate. By proposing that societal systems of provision be examined, redesigned and reconfigured in line with sustainable consumption goals, the New Economics proposes nothing less than a paradigm shift for the economy, or a wholesale transition in the presiding 'regime'. This implies that rather than making incremental changes, the model entails a widespread regime change for the economy and society, altering the rules of the game and the objective of economic development.

Unsurprisingly perhaps, this eclectic body of thought rejects economic individualism, and pays particular attention to the contextual – social, psychological and structural – factors which influence consumption practices. For example, whereas the mainstream approach to sustainable consumption relies on 'green consumers' playing their part in the marketplace, the New Economics instead addresses 'Ecological Citizens' who act ethically in public and in private to reconfigure the patterns of their lives to reduce environmental and social impacts on others (Dobson, 2003). The New Economics is fundamentally an equity-based understanding of environmental governance, drawing on 'ecological footprinting' metaphors to guide action. Ecological footprints define and visualise environmental injustice in terms of the inequitable distribution of 'ecological space' (the footprint of resources and pollution-absorbing capacity) taken up by individuals, cities and countries; this inequity requires a reduction in the scale of material consumption among the affluent advanced economies (Wackernagel and Rees, 1996).

Seeds of change: the New Economics in practice

This book critically assesses current mainstream policy responses to the sustainable consumption agenda, and consolidates an alternative, New Economics approach. It examines ecological citizenship at perhaps its most mundane, yet its most ubiquitous and fundamental, level: the choices and actions which individuals and households make on a daily basis, in the supermarket and on the high street. It deals with changing consumption patterns, consumer behaviour and lifestyles, and how these relate to environmental and social demands for sustainability. 'Sustainable consumption' has become a core policy objective of the new millennium in national and international arenas, despite the fact that its precise definition is as elusive as that of its companion on the environmental agenda, sustainable development. Current patterns of consumption are, quite clearly, unjust and unsustainable; the extent and nature of the transformation required is hotly debated, reflecting as it does competing deep-rooted beliefs about society and nature (Seyfang, 2004a). For some, it is sufficient to 'clean up' polluting production processes and thereby produce 'greener' products (OECD, 2002b; DEFRA, 2003b); for others, a wholesale rethinking of affluent lifestyles and material consumption *per se* is required (Douthwaite, 1992; Schumacher, 1993).

Chapters 2 and 3 examine these two positions in greater depth, highlighting the theoretical foundations of each perspective. While the mainstream approach is well-represented in policy frameworks, the New Economics perspective currently exists largely outside this world. Nevertheless it is strongly represented by networks of grassroots initiatives and community activists, many of them inspired by the Rio Summit itself, working to challenge existing practices, and create new social and economic institutions which allow people to express these ecological citizenship values in their daily lives (Church and Elster, 2002; Seyfang and Smith, 2007). Consequently, much practical New Economics work on sustainable consumption involves innovation and experimentation on a small scale, in the hope that successful practices will grow and expand, so influencing wider mainstream systems of provision. Chapter 4 (co-authored with Adrian Smith) sets out a major theoretical framework for the book, a conceptual lens through which the later empirical work is viewed. New ideas about innovation and transitions in socio-technical systems are applied to grassroots community-based experiments for sustainable development, bridging two previously unrelated areas of theory and policy, by seeing them as *innovative green niches*. We distinguish between market-based (usually technological) innovations, and community-based (usually social) innovations, and begin to explore the implications of viewing the grassroots as a neglected site of innovation for sustainable development. The challenge is to identify how niche social innovations can grow and spread into mainstream society, and to articulate a theory of change within an approach that might otherwise emphasise the constraints of social infrastructure too heavily (Smith, 2007).

There then follows a series of thematic discussions on aspects of sustainable consumption within the New Economics approach, looking at three fundamental areas of provision: food, housing and finance. Each of these chapters presents case studies of grassroots innovations which attempt to actualise the theory through practice, and brings empirical research to bear on theory through evaluative studies. Chapter 5 examines sustainable food, and reports on a local organic food cooperative which aims to provide a socially just and ecologically responsible system of food provision, bypassing supermarket distribution channels in favour of farmers' markets and directly supplying consumers. The threats posed by mainstream supermarkets seeking to attract customers interested in local and organic foods are outlined, to assess the scope for alternative initiatives like this to survive mainstream competition. Chapter 6 addresses housing provision and presents cases of innovative builders aiming to improve the sustainability of building technology and develop socially sustainable models of housing provision, but who find themselves on the margins of mainstream housing provision regimes, struggling to achieve wider influence. Here the scope for innovations to challenge and influence mainstream processes is examined. Chapter 7 turns to finance, and describes a range of complementary currencies, which are assessed for their contribution to sustainable consumption. These are alternative monetary tools which aim to overcome structural weaknesses in mainstream money, by incentivising more sustainable behaviour. The relationship between the alternative economic space, and the mainstream regime of work and income distribution is discussed.

Finally, the book concludes with an assessment of the state of the New Economics of sustainable consumption in theory and in practice, and its outlook for the future. It reflects on the empirical discussions, and applies transitions management theories to reveal common experiences across all three areas in terms of niche-regime dynamics. These suggest that bottom-up New Economics initiatives do have the potential to influence wider society, but their oppositional framing means that they fail to resonate strongly with the mainstream regime, preventing the successful translation of ideas. They also require top-down support and policy space, in which to grow and thrive. Measures to address this failure are discussed, drawing on existing knowledge of innovation systems and applying them to this new context, but there is much work to be done. A new policy and research agenda is presented to enable the innovative potential of grassroots innovations to be harnessed for sustainable development.

2
Sustainable Consumption:
A Mainstream Agenda

> Choosing a car is one of the most environmentally-sensitive decisions you can make... Motoring on the Green Consumer Guide features the most environmentally-sound cars available in the UK today.
>
> (Green Consumer Guide.com, 2007)

> We cannot permit the extreme in the environmental movement to shut down the United States. We cannot shut down the lives of many Americans by going extreme on the environment.
>
> (President George Bush Sr at UNCED, quoted in
> *The Guardian*, June 1 1992)

Green has gone mainstream. Between 2002 and 2006 the UK retail market for ethical goods and services grew by over 80% to £32.3 billion, representing an average household spend of £664 (Co-operative Bank, 2007), and mainstream media promotes the new orthodoxy of demonstrating one's ecological credentials through consumer purchases. Yet only 20 years ago this sort of lifestyle activism was the preserve of a small minority of radicals, and the mainstream economy was untouched by environmental or social concerns. How did this shift happen? How did sustainable consumption move from the margins to become the powerhouse of political change its advocates claim it represents? In this chapter we consider the institutional development of the concept of 'sustainable consumption' through a brief review of the landmark events and publications which have put sustainable consumption onto the international agenda and defined its use.

Sustainable consumption: development of an agenda

The term 'sustainable consumption' emerged in the 1990s, but one of the earliest landmark publications on the subject came 20 years earlier, in the Club of Rome's report entitled 'The Limits to Growth' (Meadows *et al.*, 1972). This Malthusian treatise against over-population (the root cause) and rising resource use (the effect) predicted famine and environmental collapse if current trends continued. It conceived of consumption entirely as a function of level of development, disregarding any social element in either its construction or maintenance: in the models, traditional consumption patterns were simply extrapolated, and the model assumed that 'as a population becomes more wealthy, it tends to consume more resources per person per year' (Meadows *et al.*, 1972: 113), in other words societies and individuals *inevitably* increase their material needs as they develop. Although 'Limits to Growth' pre-dated the 1970s oil shocks by only a few years, the overall trend proved to be falling prices for natural resources, and its predictions of catastrophe, of course, did not come to pass. However, in its focus on population growth – in the developing world – as the cause of resource scarcity, it was typical of institutional scientific and policy thinking at the time which thereby avoided questioning lifestyles and consumption in affluent nations (Cohen, 2001).

By the time 'sustainable development' emerged on the international policy agenda in 1987, a change of emphasis was becoming apparent. The World Commission on Environment and Development is best known for popularising the term with the now widely accepted definition: 'development that meets the needs of the present without compromising the ability of future generations to meet their needs' (WCED, 1987: 43). Its report was concerned with the effects of increasing material consumption to service affluent lifestyles, upon developing countries in particular, who aspire to match the consumption patterns of the west, at the expense of local environmental quality. By acknowledging this disparity of responsibility for environmental pressures between developed and developing nations, the Commission made an important step forward in highlighting lifestyles and consumption as an equity issue. They state that 'Sustainable global development requires that those who are more affluent adopt lifestyles within the planet's ecological means' (WCED, 1987: 9). Furthermore, there is a formal recognition that consumption levels are a social construct rather than an objective necessity:

> Perceived needs are socially and culturally determined, and sustainable development requires the promotion of values that encourage consumption standards that are within the bounds of the ecologically possible and to which we can all aspire (WCED, 1987: 44).

Yet despite these calls for lifestyle changes in developed countries, the report simultaneously states that continued economic growth and rising consumption is a prerequisite – or even a driver – for sustainable development, implying that efforts to cut growth in developed countries would hamper sustainability.

The 1992 United Nations Conference on Environment and Development, better known as the Rio Earth Summit, brought together world leaders and civil society groups to establish an action plan for sustainable development: Agenda 21 (UNCED, 1992). Central to the action plan is the belief that:

> the major cause of the continued deterioration of the global environment is the unsustainable pattern of consumption and production, particularly in industrialised countries, which is a matter of grave concern, aggravating poverty and imbalances (UNCED, 1992: section 4.3).

For the first time in international environmental discourse, overconsumption in the North was identified as the prime cause of unsustainable development. Chapter 4 of Agenda 21 addresses sustainable consumption and production, with a subsection explicitly on changing consumption patterns; principal among these is a concern for the unequal distribution of resource use between the affluent nations and the poorer countries (UNCED 1992, section 4.5). Agenda 21 proposes that governments should develop:

> new concepts of wealth and prosperity which allow higher standards of living through changed lifestyles and are less dependent on the Earth's finite resources and more in harmony with the Earth's carrying capacity ... this will require reorientation of existing

production and consumption patterns that have developed in industrial societies and are in turn emulated in much of the world (section 4.11, 4.15).

Clearly these sentiments challenge current developed country life-styles and aim to redefine progress in order to favour quality of life over material consumption, thereby allowing environmental 'space' for developing countries to increase their consumption levels. The plan also acknowledges the social and cultural forces driving behaviour, with the imperative to redefine social values of wealth and progress, setting new indicators and milestones for development.

But overwhelmingly, the tools identified for these transformations (including raising efficiency in production, economic instruments such as environmental taxation and eco-labelling, technology transfer, and reinforcing values that encourage sustainable production and consumption) are top-down 'ecological modernisation' strategies. These embody an environmental management strategy which 'assumes that existing political, economic, and social institutions can internalize the care for the environment' (Hajer, 1995: 25), and which aims to address the market failures which are seen as the root cause of environmental problems, and incorporate ecological factors into markets through pricing and taxation. Governments are expected to maintain a light-touch regulatory regime, allowing market signals to achieve the required changes in consumer behaviour – indeed, the 'green consumer' is a key actor in this model, for interpreting environmental information in markets, and sending 'green' signals back to producers (Hajer, 1995). Sustainable consumption in this view amounts to the consumption of more sustainably-produced goods through increased efficiency in production, economic instruments to discourage the most polluting technologies and techniques, and provision of consumer choice for greener products in the market. It advocates 'changing consumption' rather than 'reducing consumption' and falls comfortably within the scope of current political ideologies. This approach, based on a model of rational individualism, is the foundation of much of the mainstream thought on sustainable consumption, as the discussion below reveals.

Following Rio, the UN Commission for Sustainable Development (CSD) instigated a programme of work on sustainable production and consumption, and after 'Rio +5' in 1997, the United Nations

Environment Programme (UNEP) took the lead in progressing the agenda to inform the development of policies towards meeting the demands of Agenda 21 Chapter 4. Their first major report 'Consumption Opportunities' (UNEP, 2001) brings together a range of perspectives on sustainable consumption as a guidance document for policymakers, businesses and civil society in general. It focuses on approaches to 'selling' sustainable consumption to stakeholders, using information and marketing techniques to persuade people to consume differently (ethically as well as efficiently). The following year, the UN-led World Summit on Sustainable Development highlighted the goal of changing patterns of consumption and production as one of three core issues for sustainable development (UN, 2002), and its Johannesburg Plan of Implementation called for a ten-year work programme at the national and regional level to help foster the required shifts (UN, 2002), pursued through the so-called Marrakech Process of international meetings (UN Division for Sustainable Development, 2007). However, in contrast with Agenda 21 which spoke frankly about lifestyle issues, its emphasis is again on economistic, efficiency-focused instruments and policies, and the use of technology to help us consume more efficiently.

Similarly, the economistic approach is the focus of the OECD's programme of research on sustainable production and consumption launched in the wake of Agenda 21. Their definition of sustainable consumption was set out in 1994 at the OECD Symposium on Sustainable Consumption, 19–20 January, Oslo, Norway:

> the use of goods and related products which respond to basic needs and bring a better quality of life, while minimising the use of natural resources and toxic materials as well as the emissions of waste and pollutants over the life cycle, so as not to jeopardise the needs of future generations (Norwegian Ministry of Environment, 1994, cited in OECD (2002a: 9).

In 1995 a three-year work programme was launched to establish basic concepts, boundaries and frameworks of sustainable consumption, with an explicit bias towards economic analysis and tools – there is a strong emphasis on markets and economic instruments, and a diagnosis of market failure behind unsustainability (OECD, 2002b). The policies suggested to promote sustainable consumption address demand and supply-side factors which influence both 'software' or how consumers *think* and *feel* (economic and social instruments, eco-taxes and public awareness campaigns) and 'hardware', what they *do* (regulatory instruments, correcting markets, influence producers, providing a choice for consumers, eco-efficiency). This again demonstrates an ecological modernisation market-based approach to consuming more efficiently, rather than changing lifestyles.

Evidence that the scientific community was rethinking its approach to sustainable consumption came in 2000 with a collaboration between two of the most well-respected independent scientific organisations of the UK (The Royal Society) and the USA (the National Academy of Sciences) in 'Towards Sustainable Consumption'. It begins by asserting that scientific advances alone will not deliver sustainable development, without a concomitant change in human behaviour:

> It has often been assumed that population growth is the dominant problem we face. But ...We must tackle population and consumption together... For the poorer countries of the world, improved quality of life requires increased consumption of at least some essential resources. For this to be possible in the long run, the consumption patterns of the richer countries may have to change; and for global patterns of consumption to be sustainable, they must change (Heap and Kent, 2000, appendix B: 151).

The multi-disciplinary contributors to this volume seek to develop social scientific understanding of consumption patterns, and to use this to enhance the effectiveness of technical scientific contributions to sustainability.

The European Union identifies sustainable consumption and production as a key challenge within its renewed Sustainable Development Strategy (European Council, 2006), and aims to 'promote sustainable consumption and production by addressing social and economic development within the carrying capacity of ecosystems and decoupling economic growth from environmental degradation' (p. 12). Specific objectives include increasing efficiency and social performance of production, and encouraging the uptake of these products by industry and consumers, principally through better labelling and information, a continuation of a range of existing EU-wide policies and strategies (European Commission, 2004); an

EU sustainable production and consumption Action Plan is to be delivered in 2008.

Several years ahead of the EU, the UK Government launched 'Changing Patterns', its strategy for sustainable consumption and production (DEFRA, 2003b). It defines sustainable consumption and production as:

> Continuous economic and social progress that respects the limits of the Earth's ecosystems, and meets the needs and aspirations of everyone for a better quality of life, now and for future generations to come (DEFRA, 2003b: 10).

In practice, this emphasises breaking the link between consumption and environmental damage, to be achieved through an ecological modernisation strategy of market-based measures which seek to account for externalities and fill information gaps: making the polluter pay, eco-taxes, government purchasing initiatives, consumer education campaigns and instituting voluntary eco-labelling schemes. This directly builds upon the government's approach to sustainable development, which has been founded on a belief that 'stable and continued economic growth' is compatible with effective environmental protection and responsible use of natural resources: 'abandoning economic growth is not a sustainable development option' (DETR 1999: para 3, 12). It promotes 'cleaner growth' and improved resource productivity. However, the 2005 Sustainable Development Strategy 'Securing The Future' appears to have receded from this economic imperative somewhat. Instead, it talks of achieving a 'strong, stable and sustainable economy which provides prosperity and opportunities for all' (HM Government, 2005: 16) while living within environmental limits and ensuring efficient resource use is incentivised. For the first time in UK sustainable development policy, real environmental limits are acknowledged, and the metaphor of 'one planet economy' is introduced to highlight the current unsustainable developed worlds' economies which – if extrapolated to the rest of the world's populations – use the equivalent of three planets' resources (WWF, 2006).

The UK's strategy for sustainable consumption states that 'government regulation has a clear and vital role to play in ensuring that markets operate efficiently, excessive or unnecessary regulation can obstruct efficient functioning of the market' (DEFRA 2003b: 24). The government's role is therefore to correct prices and provide regulatory frameworks to influence producers to be more eco-efficient and offer consumer choices of 'green' and 'ethical' products. Hence, sustainable consumption is implicitly defined as the consumption of more ethic- ally or efficiently-produced goods (i.e. with no absolute reduction in consumption), and consumer behaviour is the driving force for change and 'market transformation' as consumers exercise their preferences for environmental goods or social rights in the market (Pearson and Seyfang, 2001; ETI, 2003b). To this end, several government-led initia- tives have begun to explore what this might mean in practice (DEFRA, 2007c). The UK's Sustainable Consumption Roundtable produced a report 'I Will If You Will' cautiously welcoming government initia- tives, but calling for government to act more boldly in driving market change, and for greater efforts to address questions of lifestyles, collec- tive action, and the economic, institutional, social and psychological barriers preventing people from consuming more sustainably (Sustain- able Consumption Roundtable, 2006).

There is evidence that these issues are starting to be taken up by policymakers. Whereas early efforts to change public behaviour relied upon a cognitive, information-deficit approach (providing better information for consumers to base their decisions on), more recent policy frameworks embrace a 'social marketing' strategy instead (attempting to 'sell' green lifestyle actions to targeted groups in society). The framework currently adopted by the UK government 'divides the public into seven clusters each sharing a distinct set of attitudes and beliefs towards the environment, environmental issues and behaviours.... [and] plots each segment against their relative willingness and ability to act' (DEFRA, 2007a: 8) in order to identify the social segments with the greatest motivation and potential to respond positively to particular behaviour-change messages. A fur- ther broadening of the UK policy framework to incorporate more consideration of contextual and social factors is demonstrated in the 2005 Sustainable Development Strategy. The new framework for policy action depicts a multi-faceted approach to stimulating more sustainable development, through actions to *exemplify, enable, engage and encourage* change, and thereby *catalyse* societal transformation (HM Government, 2005). In practice, this equates to government leading by example, provide information and facilitating structures and processes for people to help them to act more sustainably, forging partnerships and deliberative forums to learn from communities and the issues that concern them, and offering accessible, relevant channels to harness the energy and commitment of people wishing to make changes, and finally providing incentives through taxation and regulation to reward more sustainable behaviour. As a combined approach, it addresses many of the failings of previous strategies which overlooked the social contexts within which people act – for example by encouraging collective action, the sense of disempowerment felt by individuals acting alone is overcome. Nevertheless, its goal remains to encourage a leaner, more efficient consumer economy, and so it is firmly rooted in reformist ecological modernisation.

Failures of the mainstream agenda

The objectives and mechanisms of mainstream sustainable consumption policy have been discussed. This ecological modernisation approach is founded on a rational, economistic model of consumer behaviour, and assumes that consumers know and care about the social and environmental implications of their consumption habits, and have the motivation and opportunity to act on that knowledge to change their behaviour – in other words, to behave as ecological citizens when they make purchasing decisions. Conceptually, it is principally founded on the cognitive approach to understanding what drives consumption, and has in recent years begun to address some of the richer contextual, social-psychological aspects of consumption through social marketing initiatives. Furthermore, it assumes that messages sent to producers through the market have their intended effect in terms of transforming production practices. In the remainder of this chapter, a series of criticisms of this model are presented, which undermine its logic and question its appropriateness as a strategy for changing consumer behaviour. This critique progresses through issues of market functioning, measurement, assumptions and ultimately, rationales for consumption, and the goals of economic development.

Pricing failures

The first issue concerns the efficiency of the market mechanism itself. The ecological modernisation approach to sustainable consumption is predicated upon a free market system where rational self-interested actors aim to maximise utility (satisfaction). The present economic system is abstracted to a neo-classical model of limitless frontiers, insatiable wants, and optimal resource allocation through the market. Crucially, this system will only work to protect the environment when full social and ecological costs are included in market prices, yet it externalises (i.e. does not account for) the environment and society, and so sends producers and consumers the wrong signals. For example, fuel prices do not account for the costs of climate change caused by the resulting carbon emissions, and indeed aviation fuel is subsidised further as it is not taxed at all. In essence, the environment (and society) is unpriced and so value-less in the economic market – a free good to be exploited to the maximum – with severe consequences (Princen, 2002b; Daly and Cobb, 1990). Indeed, climate change is the 'greatest example of market failure we have ever seen' (Stern, 2007, p. 1).

One study examining the externalities of agriculture in the USA found that the negative impacts of crop and livestock production onto water, air, soil, wildlife and human health amounted to $5.7–$16.9 billion (£3.3–£9.7 billion) a year at a conservative estimate (Tegtmeier and Duffy, 2004). In the UK, Pretty *et al.* (2005) estimate the externalised costs (not including subsidies) of a typical weekly per capita food basket (costing £24.79) to be to £1.98, totalling £5.04 billion a year. In contrast with the much-vaunted cheapness of food available in supermarket, this cost is borne by consumer, both implicitly through declining ecological services, and explicitly through regulation, taxation and cleaning-up activities. This unwitting subsidy that the environment makes to the economy ensures that particular activities, such as current industrial agricultural practices, or transporting food around the world by air freight, or maintaining a transport infrastructure geared for private motor cars, appears economically rational despite its attendant costs (Pretty, 2001).

The UK strategy for sustainable consumption and production does recognise the externality problem and indicates some areas where full-cost pricing is being introduced, for example through the landfill tax or climate change levy on energy (DEFRA, 2003b). However, even these measures are politically fraught and easily derailed, with unpredictable social impacts as shown by the disruptive fuel protests in the UK, triggered by a small tax increase for petrol and diesel in

2000. Furthermore, by failing to put a premium on carbon throughout the market, one of the key externalities of climate change remains unaddressed (Stern, 2007). Until the full social and environmental costs of economic activity are internalised, a market mechanism will inevitably send the wrong price signals to both producers and consumers, resulting in sub-optimal efficiency and 'rational' over-consumption of resources and public goods.

Information failures

The second problematic area of the mainstream approach to sustainable consumption concerns information failures. Neo-classical economics, on which this policy is based, is predicated on the assumption that consumers are fully informed, rational, selfish, and weigh up all their options before choosing to make a purchase (Lipsey and Harbury, 1992). Governments aim to support this demand-side driver for change through a range of consumer information tools such as awareness-raising campaigns and various certification and labelling schemes which indicate the environmental and/or social performance of a product on the supermarket shelves. These labels include the Fairtrade mark for providing sustainable livelihoods to producers, Soil Association organic standards which indicate limited use of pesticides and fertilisers, the Forest Stewardship Council trademark for sustainable forestry, the European Energy Label which rates the efficiency of consumer appliances such as fridges and washing machines, and the European Ecolabel (the flower symbol) which covers a wide range of product life-cycle impacts (DEFRA, 2007b; European Commission, 2004). But despite these efforts to improve consumer awareness and support environmental decision-making, consumers still face several information barriers before they can act on their preferences in the market. First, the products they are considering may not be subject to social or eco-labelling; the EU Ecolabel has been operating since 1992 years and due to the complexity of developing and applying full life-cycle standards, is still slowly increasing its sectoral coverage (European Commission, 2004). Second, the credibility and consistency of sustainability labels is a key problem, as the plethora of labels and standards can be confusing, with unsubstantiated corporate green claims sitting shoulder to shoulder with rigorous multi-stakeholder certifications in the supermarket (Holdsworth, 2003). Previous trends towards green consumerism were derailed partly because of a perceived lack of credibility in green claims, and so simple, clear, authoritative and consistent labelling across sectors is needed (Childs and Whiting, 1998).

In addition to these market instruments, governments instigate public awareness campaigns aim to educate consumers about the impacts of their consumption and provoke behaviour change, from early 'wake up to what you can do for the environment' exhortations, to more recent efforts to promote more fuel-efficient driving habits (DoE, 1993; DfT, 2007). However recent research indicates that the 'top-down' delivery of expert information is not well received by the public – the truth, reliability and credibility of the source of the information is consistently brought into question by individuals receiving it (Hobson, 2002; see also Burgess *et al.*, 2003). Furthermore, environmental information does not simply flow from experts to lay people, rather it is processed and questioned and filtered by everyday experience to produce a complex two-way interactive production of knowledge – quite different to the learning process presupposed by the experts. Hobson concludes that the way individuals 'think about and address changing their lifestyles, and how they consider the current framing of the environmental problematique, all contrast markedly with the prevailing positivist assumptions underlying policy strategies' (Hobson, 2002: 205). This suggests that the basis of information campaigning is inadequate for its purpose, and deserves greater consideration of both its content and its context, in order to be more effective.

Self-regulation failure

The market transformation which results from effective sustainable consumerism, according to the mainstream approach, is demand-driven, as firms respond to and capitalise on the market of green and ethical consumers. However, recent experience suggests that in the case of both green and ethical consumption, most corporations only responded to public pressure when their reputations or sales were at stake, thanks to activist groups such as Corporate Watch and Ethical Consumer. While consumer demand may be the carrot, it is high-profile and potentially damaging media reports into the less palatable aspects of firms' activities which provide the very necessary stick to prompt changes in corporate behaviour (Pearson and Seyfang, 2001). Even these voluntary changes are vulnerable to erosion and shifting trends and rather than making continual progress, reversals in corporate policy are not uncommon. In the UK, Little-woods clothing stores were a major participant in the Ethical Trading Initiative (ETI), but a change of management led to its withdrawal from the ETI and its ethical trading team being closed down, as corporate responsibility was not seen as an important issue to consumers (ETI, 2003a). Green consumerism was a growing trend during the early 1990s, but as a result of changes in consumer preference during the 1990s, sales of 'green' product ranges fell and many supermarket own-brand ranges of 'green' cleaning products, for example, were discontinued (Childs and Whiting, 1998). These examples suggest that the social or environmental improvements made as a response to consumer pressure have been rescinded as attention shifted, rather than taken up as new minimum standards, and that 'left to their own devices, [transnational corporations] are likely to fulfil their responsibilities in a minimalist and fragmentary fashion … they still need strong and effective regulation and a coherent response from civil society' (UNRISD, 2000: 90). As a result, the market-based mechanism of voluntary self-regulation cannot be relied upon to deliver continual progress towards more sustainable production and consumption, as firms behave in a (rationally) opportunistic fashion, taking advantage of the lack of government regulation, to advance and recede as consumer attention demands.

Measurement failure

It is a truism that what gets counted, counts. The key economic indicator used by governments the world over is gross domestic product (GDP), a measure of economic activity which is often used to represent consumption (and by implication wealth and wellbeing). Critics have argued it is simply the wrong indicator to use (Anderson, 1991; Douthwaite, 1992; Jackson, 2004a). This measure makes no distinction between those activities which represent enhancements to quality of life, and those which do not (expenditure on pollution clean-ups, for instance). This results in economic policy designed to increase GDP through targets of continual economic growth, which commonly translate into sustainable development strategies as a given (DETR, 1999). The UK's Commission on Sustainable Development argued that sustainable development must not be linked to economic growth, as the government is wont to do, as this 't[ies] it to the very economic framework that was responsible for unsustainable development' (SDC, 2001: para. 32), and indeed this prerequisite for continued economic growth has been removed from the most recent UK sustainable development strategy (HM Government, 2005).

The purpose of economic activity, for economists, is 'maximising utility', which may also be described as 'enhancing quality of life', which in turn is related to factors such as basic human needs for shelter and food, clothing and water, plus social needs like companionship, belonging to a community, and freedom of spiritual practice. There are further determinants of quality of life, such as good health, rest and recreation time, fulfilling work, community and cultural participation, and opportunities for personal development (UNEP, 2001). Of course, many of these do not require material consumption at all, and UNEP argues that increasing consumption levels are contributing to lower quality of life through overwork, degraded environments and social breakdown. So if the ultimate aim of economic activity is to improve quality of life, then material consumption levels *per se* can no longer be considered an adequate proxy for this. Indeed recent research points to the fact that while GDP has continued to rise for the last 20 years or so in developed countries, measures of wellbeing have remained relatively stable (Marks *et al.*, 2006; Thompson *et al.*, 2007; Jackson and Marks, 1999). While the UK government has adopted sets of social and environmental indicators – including in 2007 a series of personal wellbeing indicators – as part of its sustainable development strategy (DEFRA, 2007d), these are 'add-ons' to the principal economic indicators. Until more accurate measures of wellbeing are fundamentally accepted by governments, sustainable consumption strategies, and development policies in general, will, by definition, be off-target.

Enfranchisement failure

A common analogy made of sustainable consumption is that it is a modern form of political citizenship, of making one's preferences known, taking action on the basis of those values with the intention of changing social and environmental conditions in society, and essentially *voting* with one's money. In a relatively early example of this framing, Zadek *et al.* (1998) write about 'purchasing power' as a form of civil action (and a complement to an earlier generation of activism based on boycotts), and assert that consumption choices 'can change the manner in which business is done and the terms by which livelihoods are constructed' (p. 1) by directly benefiting stakeholders, influencing larger-scale processes, and combating passivity by demonstrating that positive alternatives are possible. More recently Barnett *et al.* (2005) claim that rather than democratic participation being in decline, ethical consumerism is a new terrain of political action which links individual action with collective outcomes, through governance of consumption patterns and indeed the consumer themselves. Concurring with these findings, Shaw *et al.* (2006) find that ethical consumers perceive their actions as 'empowering' and think in terms of a voting metaphor. Yet while these analyses are initially encouraging in terms of realising the potential of citizenship actions, a fundamental contradiction lies at their heart: namely, that it is a citizenship of the *market*, and *individual consumption* purchases are the only votes that count.

Within this framework of political action and market transformation, there are many barriers which may prevent individuals from acting on their ecological citizenship preferences, leaving them unable to influence the market. These include the affordability, availability and convenience of more sustainable products and services; feelings of powerlessness generated by the thought that individual action will not make any difference; and disenchantment with corporate green marketing (Holdsworth, 2003; Bibbings, 2004). One barrier to effectiveness is that 'institutional consumption' decisions are made on a societal level, rather than by individuals, and only products and brands with which consumers are familiar are subject to transformative consumer pressure. Institutional consumption, which includes producer goods, public procurement (purchasing by the state for building and maintaining roads, hospitals, schools, the military, and so forth, accounts for half of all consumption throughout western Europe) and most investment products, is extraneous to the hands of individual domestic consumers, according to Lodziak (2002). Consequently the majority of societal consumption takes place in a consumption decision-making arena beyond the reach of individuals, and so excluded from the market transformation possibilities of sustainable consumption.

Another important barrier is a consumer preference for products that are simply not available, or for avoiding or reduced consumption in the first place – a choice *not* to consume is as meaning-filled as one to consume, yet makes no impact on the market. Princen (2002a) argues this may be because the mainstream neo-classical economic model on which policy is based cannot account for activities and transactions which take place outside the market, yet for consumers the choice to seek 'less consumptive, less material-intensive means of satisfying a need' (p. 28) can be the strongest expression of sustainable consumerism. Consequently, if we stick with the voting metaphor, then the sustainable consumption marketplace begins to look like a rigged election, disenfranchising those who cannot afford their ballot papers, or who prefer to reduce consumption or choose a collective alternative to individualised market choices, voting for 'none of the above'!

Equity failure

A further factor which limits the effectiveness of a mainstream sustainable consumption approach is the category error which pits individuals against global institutions to solve global collective action problems. Sustainable consumption as defined in mainstream policy relies upon the summation of many small acts of atomised consumer sovereignty to shift the market. However, the environmental problems which this strategy seeks to address, such as climate change, are global in nature (crossing state boundaries, and distributing risks, benefits and costs unevenly among stakeholders with vastly different capacities to respond) and require negotiated, collective efforts to resolve. Furthermore, the institutions which currently propagate unsustainable consumption are also global, such as the World Trade Organization whose rules prevent governments favouring fairly-traded or 'green' imports (Tallontire and Blowfield, 2000). Transforming these institutions to serve ecological citizenship requires collective strategic action (Manno, 2002). While 'green growth' and 'market transformation' offer the promise of an environmentally friendly future which does not threaten the political or commercial status quo, green consumerism and individualisation of responsibility for the environment belie the powerful institutions and interests at stake. Maniates (2002) states that 'when responsibility for environmental problems is individualised, there is little room to ponder institutions, the nature and exercise of political power, or ways of collectively changing the distribution of power and influence in society' (p. 45). Indeed, the institutional consumption in society (by governments, primarily on social infrastructure, the military, etc) which makes up the bulk of resource use is collectively determined, albeit somewhat invisibly, yet policy attention focuses on the behavioural responsibilities of the individual and household. Thus the mainstream strategy lacks political bite, by placing relatively powerless individuals against institutional behemoths, and neglects governance issues and questions of power.

The hidden mainstream agenda

A historical review reveals that from its auspicious origins at Rio, the term 'sustainable consumption' has evolved through a range of international policy arenas, and its definition narrowed as it became more widely accepted as a policy goal. More challenging ideas became marginalised as governments instead focused on politically acceptable and economically rational tools for changing consumption patterns such as cleaning up production processes and marketing green products – instruments and approaches that fit well within current styles of governance. So the agenda has shrunk from initial possibilities of redefining prosperity and wealth and radically transforming lifestyles, to a focus on improving resource productivity and marketing 'green' or 'ethical' products such as fairly traded coffee, low-energy light bulbs, more fuel-efficient vehicles, biodegradable washing powder, and so forth.

As Jackson asserts: 'Reasons for this institutional retreat from the thorny issues of consumer behaviour and lifestyles are not particularly hard to find' (Jackson, 2007a: 6). Policy intervention to regulate the aspirations and choices of individuals goes against the political grain, and contravenes myths of consumer sovereignty, not to mention threatening vested interests in current consumption patterns. Indeed, it has been argued that the green consumer approach to sustainable consumption is more about advertising one's values and sense of style, than about transforming markets. Steffen (2007) argues that:

the vast majority of the green products around us are, at best, a form of advertisement for the idea that we should live sustainably, a sort of shopping therapy for the ecologically guilty... Even worse, the glut of green shopping opportunities is overshadowing the most basic message of all, which is that the most sustainable product is the one you never bought in the first place.

From this critical perspective then, green consumerism and the mainstream strategy which underlies it is selling one crucial meta-message: namely that consumerism as a way of life offers answers to our problems, and is not negotiable.

The latter part of this chapter has assessed the scope and potential of the mainstream policy model of sustainable consumption, and has found it to be limited by a number of factors. These relate to the efficiency of the market mechanism, and the ability of that mechanism to effectively deliver the outcomes ecological citizens desire. Despite these limitations, it would be premature to dismiss the mainstream model of sustainable consumption entirely, not least because of its ubiquity and apparent political acceptability (though of course these may be inversely related to its ability to challenge current institutions!). The approach does appear to achieve significant benefits in terms of raising awareness of the social and environmental impacts of behaviour, and encouraging individuals to think about these and reflect upon the difference they can make through their consumption patterns. It may be that this process is the first step on a journey towards greater education and activism concerning ecological citizenship, and there is certainly scope for strengthening the improvements which mainstream sustainable consumption can make – namely, through government regulation, addressing market failures, and supporting firms in transforming their activities.

It is clear that mainstream sustainable consumption, as practised in policy arenas and embodied in the UK's strategy for sustainable consumption and production, is a limited tool for change, but may nevertheless be a useful stepping stone on the way, for firms and consumers alike. In the next chapter we examine an alternative approach to sustainable consumption which aims to address the failings outlined here.

3
Sustainable Consumption and the New Economics

Despite the direction the mainstream policy framework for sustainable consumption has taken, the challenge laid down at Rio has not fallen on deaf ears. To recall, that objective was not only to promote greater efficiency in resource use, but also to realign development goals according to wider social and environmental priorities rather than narrow economic criteria, and to consider the possibilities of lifestyles founded upon values other than consumerism. This alternative approach to environmental governance and sustainable consumption is supported by a broad body of thought known collectively as the 'New Economics', which is elaborated in this chapter. It is founded on new conceptions of wealth and work, new uses of money, and an integral ethical stance; when it comes to consumption issues, it embodies what Jackson (2004b) terms an ecological critique of the utilitarian approach to understanding consumer motivation. Consuming more, simply put, does not necessarily make us happy, healthy, wealthy or wise. This view, for a long time considered taboo in policy-making circles, is finally starting to be heard in mainstream forums. The challenge for this broad church of interdisciplinary and alternative perspectives – and this chapter – is to provide a coherent theoretical foundation for policy and action.

Evolution of an alternative agenda

The New Economics is a philosophical and political school of thought founded on a belief that economics cannot be divorced from its foundations in environmental and social contexts. Although its roots go back to twin traditions of environmentalism and social economics (see Pepper (1996) and Lutz (1999) for excellent reviews), it has emerged in recent years from the environmental movement and built upon the work of writers such as E. F. Schumacher (1993) to develop a body of theory about how a 'humanistic' economics concerned with justice and social wellbeing could be envisioned and practised. Schumacher's landmark book 'Small is Beautiful' proposed a human-centred alternative to mainstream neo-classical economics in which social and environmental wealth is valued and protected within the context of 'human-scale' participatory democracy, localised economies and modest consumption levels. He termed this 'Buddhist economics', an 'economics as if people mattered' and this preoccupation with scale – the scale of the economy and the scale of social organisations which best serve the development of human potential – resonated with other writers of the 1970s seeking answers to the international political and economic disruptions of the oil crises, the cold war, and the need to live within environmental limits. Other notable writers in this field include Lutz and Lux (1988) on a humanistic economics, Sale (1980) on 'Human Scale' economies, Meadows *et al.* (1972) on the environmental 'Limits to Growth', Boulding (1966) on the 'Spaceship Earth' concept which pictured the planet as a closed ecological system, rather than the limitless frontier favoured by mainstream economists. Røpke and Reisch succinctly identify the role of consumption studies within 'ecological economics':

> The scale of the human economy in relation to natural systems is now so large that basic life support systems for humans are threatened. As the continuous growth of the economy's scale ought to be curbed, it is not possible to rely on economic growth to solve the global problems of poverty. [...] In other words, to improve the environmental space for increasing living standards for the poor, the rich have, at least, to stop the increase in their appropriation of natural resources and pollution absorption capacity (Røpke and Reisch, 2004: 5).

Inherent in this position is a concern for equity between and across generations, and distributional issues in the allocation of this environmental 'space'. The ethical stance underpinning New Economics is elaborated below, using Dobson's 'ecological citizenship' model of an active citizen engaged in changing behaviour at the individual and collective level, in public and in private (Dobson, 2003). What these 'deep green' thinkers share is a rejection of mainstream 'light green' approaches to the environment which presume an incrementally-improved 'business as usual' approach to sustainable development. Instead, they hold a conception of a sustainable future which include radical re-organising of economies to be more localised, decentralised, smaller-scale, and oriented towards human wellbeing, equity, justice and environmental protection. Furthermore, the political prescriptions of these normative analyses – in direct contradiction to mainstream policies – lend themselves to supporting a growing movement of academics and activists seeking change (Ekins, 1992; Dauncey, 1996).

The term 'New Economics' was first adopted in 1984 following a gathering of these alternative thinkers in a parallel conference to the high-profile G7 summit of the seven richest industrial nations. Known as 'The Other Economic Summit' or TOES, this event focused on what was termed 'real-life' economics and addressed subjects such as international debt, local economic resilience, valuing the environment, building social cohesion and so on, through new theoretical frameworks and nascent demonstrations of these principles in practice, such as Local Exchange Trading Schemes as described in Chapter 7 (Ekins, 1986). A direct consequence of the successful TOES was the establishment in 1986 of the UK's New Economics Foundation (NEF), a charitable organisation (and now a self-styled 'think-and-do-tank') with the aim of further developing ideas and practices of 'economics as if people and the planet mattered' and influencing policy (see www.neweconomics.org). So what are the central tenets of New Economics? Boyle's pamphlet 'What is New Economics' explains:

> Old economics loses everything it fails to measure. Women are lost if economics simply tracks what happens to households as a whole. Nature is lost, because its services are not valued. People working in the informal economy – up to a third of people in poor countries – are lost. The contribution of traditional knowledge and cultural diversity are lost. [...] Since the first days of TOES, new schools of economic study have been sprung up to study what old economics misses out: how organisations work

(institutional economics), the contribution of nature (ecological economics) and human behaviour (socio-economics). This broader approach gives us new insights, new policies, but also different assumption to begin with. [...] This approach relies on a broader understanding of what we mean by *wealth*, a richer conception of *work*, new uses of *money*, and on integrating *ethics* back into economic life (Boyle, 1993: 5, emphasis in original).

Let us take each of these four core assumptions in turn and examine their substance and implications for sustainable consumption in practice.

A broader understanding of wealth

First, redefining 'wealth' and 'prosperity' is a key element of the New Economics, which incorporates both environmental as well as social values. The previous chapter discussed the consequences of a narrow conception of value, externalising environmental costs and undermining social cohesion. Here, those sources of value are foregrounded. Building on the lessons of ecological economics (Costanza, 1991), the New Economics places the environment at the heart of its economic analysis, accepting that there are ecological services that cannot be substituted for other types of capital, and that ecosystems do not react in a predictable, linear way to external stresses. Following from this, the economy cannot be viewed as an abstracted mechanism for indefinitely producing 'value' but rather has to take its place within the environment – and society – as a starting point. This in turn demands alternative sets of indicators which redefine 'progress' and 'wealth' to achieve a greater appreciation of 'well-being' and 'quality of life' (Ekins *et al.*, 1992; Ekins and Max-Neef, 1993), arguably a better measure of societal progress and the true objectives of economic activity, than increasing consumption as measured by conventional indicators such as GDP.

Consequently new sets of indicators of economic and social progress such as Daly and Cobb's 'Index of Sustainable Economic Welfare' (1990) and its derivative, the Measure of Domestic Progress or MDP have been proposed to better capture this wealth creation at the national level (Jackson and Marks, 1994, 1999; Jackson, 2004a; see also Anderson, 1991) and also at the local level where social capital, community spirit and engagement are also valued (Walker *et al.*, 2000). The MDP index finds that while GDP has increased rapidly since 1950, MDP has barely grown at all. The divergence is more noticeable in the last 30 years, as GDP has grown by 80% but MDP has fallen during the 1980s mainly due to environmental degradation, growing inequality and associated social costs, and has still not regained the peak achieved in 1976 (Jackson, 2004a). As this report states: 'every society clings to a myth by which it lives; ours is the myth of economic progress' (Jackson, 2004a: 1). Alternative approaches to sustainable consumption require governments and society to rethink the purpose of economies – is it to increase welfare, or to boost economic activity? – and design policies which achieve the underlying goals, rather than their proxies. Daly and Cobb (1990: 373) state 'We hope that someday measures of the health of communities could guide policymaking rather than the per capita availability of economic goods', and for the first time, in 2007 a headline 'wellbeing' indicator was introduced to the UK's national indicators for sustainable development. Here, wellbeing is defined as 'a positive physical, social and mental state' (DEFRA, 2007d: 111) and the indicator is a composite of a range of social and environmental 'goods' e.g. freedom of fear of crime, life satisfaction, life expectancy, housing conditions, access to amenities etc. To return to an assertion made earlier, what counts is what gets counted, and development goals will always be geared towards the indicators chosen. MDP is found to closely match life-satisfaction indexes, which have not risen significantly for 30 years, and so this measure might be a good example of an alternative national accounting mechanism which embodies and so promotes the values of ecological citizenship (Jackson, 2004a).

An important consequence of this principle is a recognition that continual economic growth, and increasing globalisation may not be the best way of achieving greater societal wellbeing, if it brings with it growing inequality, hidden costs, and greater vulnerability of local economies to global restructuring and external economic shocks. New Economics therefore favours the growth of decentralised social and economic organisation and local self-reliance (Jacobs, 1984; Schumacher, 1993), proposing an 'evolution from today's international economy to an ecologically sustainable, decentralizing, multi-level one-world economic system' (Robertson, 1999: 6) or what is known today as the 'new localism' (Filkin *et al.*, 2000). Most fundamentally, it proposes a 'steady state' economy, rejecting the imperative of continual economic growth as being unsustainable, and instead devoted to the growth of wellbeing – meeting needs – rather than material consumption (Daly, 1992; Henderson, 1995; Douthwaite, 1992).

On the subject of 'needs', the New Economics diverges from neoclassical economics (wherein individuals are presumed to have insatiable wants), and from Maslow's (1954) hierarchy of needs (whereby 'non-material' needs are attended to only when subsistence needs are met). Notably Max-Neef's (1992) conceptualisation of a universal, finite set of human needs, which can be met through a variety of means, provides a useful tool for understanding the New Economics approach. He identifies the following categories of needs: subsistence, protection, affection, understanding, participation, leisure, creation, identity, freedom. Each of these have four components: being, having, doing and interacting, and the resulting matrix is intended as an evaluation tool for assessing the wealths and poverties of a community. Crucially, beyond the level of subsistence, there is no hierarchy of needs attainment, but rather trade-offs, complementarity, contradictions and synergies between 'satisfiers' of various aspects of each need. Satisfiers may be pseudo-satisfiers which give a false impression of meeting needs; they may be simple singular satisfiers; they may inhibit the satisfaction of other needs, and they may synergistically satisfy several needs at once. While the needs-matrix is universal, Max-Neef stipulates that 'the way in which needs are expressed through satisfiers varies according to historical period and culture...*and economic goods are [satisfiers'] material manifestation'* (Max-Neef, 1992: 203, emphasis in original). We can take from this an approach which posits a range of alternative methods of meeting needs which do not rely on private ownership or material goods (see for example Briceno and Stagl (2006) for a discussion of Product Service Systems which attempt just this), and a possibility that the culturally and historically-specific consumerist economy is neither an inherent outcome of human needs-satisfaction, nor an inevitable one.

If the fundamental goal of development is improving wellbeing in a socially and environmentally sustainable manner, then once this underlying objective is accepted, it is possible to conceive of many ways in which this can be achieved while reducing material consumption, resource use and conventional economic activity. One group of people experimenting with notions of cutting consumption while improving wellbeing and needs-satisfaction are known as 'voluntary simplifiers'. This movement began in the 1970s in response to the emerging environmental 'limits to growth' debate as well as a growing disenchantment with materialism and an interest in personal development (Elgin, 1981; Etzioni, 1998). Adopting a lifestyle of voluntary simplicity involves choosing to consume less (perhaps entailing working less and accepting a cut in income) and embracing a less materially-intensive lifestyle, in search of greater happiness and fulfilment. In the 1990s the same principles were characterised as a 'downshifting' trend of cash-rich and time-poor professionals cutting their working hours and income in exchange for simpler living and a higher quality of life (Ghazi and Jones, 1997). Some voluntary simplifiers emphasise the environmental aspects of their lifestyles, and Simms (2003) explicitly recalls wartime rationing as a model for restraint in consumption in his model of an 'environmental war economy'. But for others it is a matter of personal fulfilment, rejecting materialism and consumerism because of its perceived detrimental effects on self-esteem and contentment (Bekin *et al.*, 2005; McDonald *et al.*, 2006).

Manno (2002), for example, describes an economy of 'care and connection' where 'non-commodity' goods are produced and exchanged. These are goods and services which embody qualities which cannot easily be mass-marketed and sold, and they are typically produced locally to the site of consumption, embodying webs of relationships, and are collectively owned. This is in contrast to highly commodifiable goods and services which are standardised, free of social relationships, mobile, convenient and with clear private ownership properties, and represent larger ecological footprints than their non-commodified alternatives. From this analysis, ecological citizens should challenge the commercial, political and legal forces which currently favour commodification, to produce instead locally significant social economies, where collective ownership and co-production take precedence.

A richer conception of work

The second departure that New Economics makes from the mainstream is in its conception of 'work'. It proposes that the economic 'lens' is extended to include the diverse categories of non-commodified labour which coexist alongside formal employment, and which includes the unpaid socially reproductive labour which sustains communities and families. This work – bringing up children, volunteering in communities, helping neighbours – must be recognised, accounted for, and protected in order to strengthen inclusive, resilient communities and so support the market economy which rests upon this bedrock, it is claimed. Henderson illustrates this principle of a more inclusive definition of economic activity through her 'layer cake' analogy. In this model of the total productive system of an industrial society, the top half of the cake represents the monetised economy which is officially measured and included in national accounts. The icing on the very top represents the private sector (investment, savings, employment, commerce), which sits upon the public sector (infrastructure, military, schools, healthcare etc). There is also a small 'underground' monetised economy of undeclared work. This monetised half of the productive system rests upon – and is subsidised by – a non-monetised lower half of the cake, which is the work, wealth and resources normally excluded from economic analysis. This is comprised of the social economy (mutual aid, self-provisioning, non-market exchange) and beneath that, at the foundation of the entire system, lies the environment as a provider of habitable living conditions, a resource, an absorber of pollution, and a waste-recycler, within limits (Henderson, 1995: 30).

This approach to viewing the economy requires a redefinition of 'work' to value the unpaid labour and the informal employment in society. It echoes Waring's landmark work 'If Women Counted' (1988) on women's unpaid labour and the parallels with unvalued environmental inputs, both contributing to gender inequality and misdirected economic development. The scale of this labour is considerable: one estimate of the replacement value of unpaid labour in Canada in 1992 was $285 billion (Dresher, 1997). The extent to which this labour is currently marginalised and discounted is demonstrated by prevalent modes of social policy within industrial societies which aim to eradicate informal employment and insert all able citizens into formal employment, thereby removing workers from the 'economically inactive' informal and voluntary sectors, and forcing a withdrawal of the unpaid work which maintains healthy communities (Henderson, 1995; Seyfang, 2004c; Hoskyns and Rai, 2007). Furthermore, the realm of economic activity is not as commodified nor as homogenous as is generally presumed in the standard economic myth of the universal market (Gibson-Graham, 1996; Leyshon *et al.*, 2003).

Williams (2005) examines the scope and scale of non-marketed work and finds that its extent is much greater than commonly assumed, and that rather than being a residual category, in some countries its share of total employment is growing over time. Participation in informal employment and other non-marketed forms of work brings many social and community benefits as well as economic advantages, including building social capital, accessing networks of support, developing skills and confidence, meeting needs and growing infant businesses.

In his book 'Future Work', Robertson (1985) proposes that a sustainable economy would allow people to have a portfolio of employment options, and undertake a variety of different forms of work – domestic labour, unpaid work in the community, informal employment for cash and local currencies, and formal employment in the market economy – which are each valuable and valued in their own right. Furthermore, by valuing and rewarding hitherto marginalised categories of work such as unpaid domestic and social reproduction, such an approach would do much to boost the esteem and confidence of its practitioners, who, as a result, may find themselves less likely to engage in consumption to meet those social and psychological needs for public recognition, self-expression and self-esteem. Schor (1991) identifies an institutionalised 'work and spend' treadmill in modern consumer societies which compels people to work more, to earn a higher income, which is then spent on consumer goods required to cope with and compensate for the long hours worked (convenience food, holidays, massages, childcare). This cycle of work and consumption ensures that productivity gains are taken in the form of higher income, rather than greater leisure time, and has enormous impacts on household consumption patterns. To the extent that the current model of 'full-time formal employment' is the norm, this could be mitigated and reframed within a context of more flexible and diverse working patterns, where individuals have the time available to meet their needs other than by consuming more, and consequently require a lower income to do so – coinciding with the downshifters and voluntary simplifiers of the previous section.

New uses of money

The third distinctive characteristic of New Economics is its understanding of money. Mainstream economics describes money as a neutral measuring tool which meets the criteria of being a means of exchange, a store of value and a unit of account (Lipsey and Harbury, 1992). New Economists claim that not only do these functions of money conflict with each other in modern use (e.g. withdrawing money from the economy to store value prevents money circulating to meet needs), but that since all money systems are socially constructed infrastructure, the design of exchange mechanisms builds in particular purposes and characteristics to each type of money, which in turn promotes particular types of behaviour. Lietaer states 'Money matters. The way money is created and administered in a given society makes a deep impression on values and relationships within that society. More specifically, the *type* of currency used in a society encourages – or discourages – specific emotions or behaviour patterns' (Lietaer, 2001: 4). Mainstream money is a tool of an economic system which prioritises a narrowly defined range of economic activities (by valuing what is scarce rather than what contributes to wellbeing), in isolation from social and environmental contexts, and so inhibits sustainable consumption. Therefore new systems of exchange need to be invented, specifically designed to serve different ends by taking a 'whole systems' approach to the economy-society-environment context of economic activity. While these may be less efficient from a purely economic viewpoint, they are actually *more* rational when one incorporates environmental and social factors into the equations (Greco, 1994; Boyle, 2002; Seyfang, 2000; Lietaer, 2001).

One such new monetary initiative which has been proposed is 'community currencies' (the subject of Chapter 7 of this book), the generic term for a wealth of contemporary alternative exchange systems which exist alongside mainstream money, and which have been springing up in developed and developing countries since the 1990s as a response to social, economic and environmental needs. Alternative money systems are not new; efforts to reform, replace and redesign money have a long and rich history around the world as a tool to support local economies in times of recession (when conventional money is worthless or in short supply), and it is only in recent decades that the notion of having an exclusive national currency became the norm (Seyfang, 2000; Tibbett, 1997; Douthwaite, 1996; Boyle, 2002). In recent times they have emerged as community responses to the economic, social and environmental pres-sures of globalisation and economic restructuring, and the social embeddedness of economic relations has become a more significant objective (Seyfang, 2001b). For example, community currencies have arisen in Mexico, Uruguay, Senegal, Thailand, Japan (DeMeulenaere, 2004), and in Argentina, alternative money systems traded in barter markets and conceived as a 'solidarity economy' by local environmentalists became real lifelines for much of the population during the national economic crisis in 2001–2 (Pearson, 2003). Sociologist Nigel Dodd (1994) proposes that the five essential characteristics of monetary networks are: accountancy, regulation, reflexivity, sociality and spatiality; and a study by Lee *et al.* (2004) maps out a range of community currencies against these criteria. This study finds that the alternative monetary networks each have those characteristics to different degrees and in different forms, and furthermore they differ from the mainstream monetary network in each of the five dimensions, suggesting that they do indeed offer something functional yet distinct to the mainstream system of provision for money.

In addition to these new *types* of money, the New Economics also proposes a range of new *uses* of mainstream currency which aim to meet the economic and social needs of individuals and communities which current financial provision overlook and exclude. These include credit unions, ecological banks and building societies, community development microfinance institutions for small-scale business start-ups, community reinvestment mechanisms, and so on (Meeker-Lowry, 1995). What these tools have in common with complementary currencies and new types of money, is that they begin to offer a framework for a new system of financial provision, based on different rules to the mainstream, which express different values and understandings of what is valuable, and how the economy should interact with society.

Integrating ethics back into economic life

Fourth, New Economics is concerned with ethics. Unlike the positive, apolitical abstractions of mainstream economics (which translate to very ideologically-based policy prescriptions) it is a normative analytical approach, which aims to describe and facilitate the transition to a more sustainable society. It therefore takes explicit moral stances about the role of government, commerce and the social economy in delivering such a world, and about what the aims of policy should be – namely increasing sustainable wellbeing while maintaining healthy ecosystems (see for example the New Economics Foundation's 'Wellbeing Manifesto for a Flourishing Society' by Shah and Marks (2004)). A tool which expresses this ethical stance, and the importance of ethical practices within New Economics, is multi-criteria evaluation methodology. One such is 'social auditing', a technique pioneered by the New Economics Foundation, for organisations and businesses to record rich accounts of social, ethical and environmental performance to be assessed alongside traditional financial accounts (Zadek *et al.*, 1997; Gonella and Pilling, 1998). The methodology departs from conventional evaluations and cost-benefit calculations by beginning with a stakeholder-defined range of objectives and indicators of success. Evidence to appraise performance is then collected and presented against each of the objectives, avoiding the need for a reductive 'bottom line', and allowing far greater transparency and accountability between organisations and their members and publics (Zadek and Evans, 1993). Initially the preserve of socially- and environmentally-motivated businesses (see for example social reports from the New Economics Foundation (1995), the Body Shop (1996) and Traidcraft (1994)), this type of ethical accounting has become far more mainstream over the last 15 years, as corporate social responsibility has developed into a big business concern – although this sometimes resulted in it being 'captured by marketing departments' (Doane, 2000: 2). Another use of these new metrics of wealth and progress is through community-based assessments of neighbourhood renewal and quality of life (Walker *et al.*, 2000; Seyfang 1999). The types of innovative evaluation methodologies pioneered through these exercises have become more widely accepted in policy-making, for example through public participation in setting and measuring sustainability indicators across a range of locally-relevant subjects, such as the number of salmon swimming in local rivers (Sustainable Seattle, 1993; Henderson, 1996; MacGillivray *et al.*, 1998; DETR, 2000).

The New Economics is an equity-based understanding of environmental governance, drawing on 'ecological footprinting' indicators to make visible global inequalities of consumption. These define and visualise environmental injustice in terms of the inequitable distribution of 'ecological space' (the footprint of resources and pollution-absorbing capacity) taken up by individuals, cities and countries; this inequity requires a reduction in the scale of material consumption among the affluent advanced economies (Wackernagel and Rees, 1996).[1] Although technically complex to calculate and the subject of debate about accuracy and methodologies, ecological footprints remain a simple, accessible way to present and understand basic issues of injustice and consumption; they are excellent communication tools. Allen remarks upon how 'one of the eye-opening consequences of living in an increasingly interdependent world is the recognition that the ties that bind people and places together often do so in unequal and unjust ways' (2006: 8). For example the Interdependence Day report illustrates how global trade ensures a web of interconnectedness across the globe between producers and consumers, and between those who enjoy the fruits of global trade and those who pay its social and environmental costs, and between those who consume more than their 'fair share' of the planet's ecological space, and those who are squeezed to the margins (Simms *et al.*, 2006). This report declares April 16th to be the UK's Interdependence Day, as it is the day in the calendar year when, if one imagined January 1 to be the start of the consuming year, UK consumers have consumed their share of the world's resources, and begin to encroach on others' ecological space. This symbolic 'ecological debt' day has moved earlier in the year over time, from July 9 in 1961, to May 14 in 1981. Furthermore, as the report attests, 'the world as a whole is also now living beyond the capacity of its ecosystems to regenerate and goes into ecological debt on the 23rd of October, causing long-term environmental degradation' (*ibid.*: 33). The political implications of this analysis are to recognise and address the inequity of such a distribution of resources, and the impossibility of the entire world consuming at the same rate as the developed countries. The New Economics therefore calls for a new 'ecological citizenship' of humanity as a whole, one which expands across borders (as does environmental change) and which recognises the political implications of private decisions and so defines everyday activities of consumption as potentially citizenly work (Dobson, 2003). The nature and characteristics of this citizenship is elaborated below.

---

[1]A related idea (technically, a sub-footprint) is the Carbon Footprint, developed to help people calculate the amount of CO_2 they emit through their energy use and transport behaviour. These are far simpler to calculate than national ecological footprints, and several online calculators exist (see for example www.carboncalculator.co.uk, and the UK government's own calculator at www.actonco2.direct.gov.uk/).

A new environmental ethic: ecological citizenship

Seeking to define and embed a new 'environmental ethic' in public debates and discourses, environmentalists aim for a rationale for changing behaviour towards more sustainable lifestyles motivated by an ethical position, rather than simply responding to superficial incentives. An environmentally informed morality implies particular types of political relationships – the nature of citizenship – between strangers, across generations and even across species (Dobson, 2003; Dobson and Valencia, 2005). This is a normative theory of change, and Dobson develops the idea of ecological citizenship by extending existing well-accepted theories of citizenship to accommodate environmental concerns, and proposes that ecological citizenship could be a motivating force for sustainable consumption.

But first, the nature of this citizenship should be described. In its traditional guises within liberalism and civic republicanism, citizenship concerns the status and activity of individuals in the public domain, in relationship to the state. Liberal political philosophy emphasises the rights of individuals, and the environment can be incorporated through a new language of environmental rights (Bell, 2005). For example, the human right to a habitable environment (as a prerequisite to all other rights) may be a sufficient claim to ensure action for sustainability. More controversially, the rights of non-human species can be argued for – challenging existing notions of who counts as a citizen – have been debated within liberalism. The second major strand of traditional citizenship thought is civic republicanism, which emphasises the duties and responsibilities that citizens have to act in the interests of the common good. Environmental responsibilities are easily introduced to this approach, as there is a great resonance with the concepts of self-sacrifice for the greater good and being an active citizen which run through green politics, encouraging people to associate the implications of their daily activities with the state of the wider environment. This dualistic notion of individuals acting according to either their personal, private interests or the collective public good is well developed within civic republicanism. Sagoff (1988) splits personal motivations into 'consumer' and 'citizen' interests, and argues that they are always in competition: the challenge is to find ways to ensure decisions are made according to 'citizen' rather than 'consumer' interests.

Citizenship is a politically contested and historically evolving term, however, and recent developments in feminism and globalisation have prompted challenges to the traditional understandings of citizenship, which have ramifications for environmentalism (Dobson, 2003). Feminism argues that the traditional constructions of citizenship are not at all universal, and are gendered and inappropriate for many women, and that the private sphere is a legitimate space for the gaze and practice of citizenship – 'the personal is political!'. When environmentalists speak of the need to change our daily actions, for example through improving energy efficiency in the home, or cycling rather than driving a car, they are describing the private sphere as a site of citizenly activity. At the same time, cosmopolitanism claims that people are citizens of all humanity rather than particular states. Clearly, this perspective resonates with environmentalism which describes us all as inhabitants of the Earth, with global environmental problems to solve which transcend state boundaries.

While clearly falling outside the traditional definitions of citizenship in terms of political status, these two challenges are based upon theories that citizenship is very much about activity, and that citizenly activity for the common good can take place at any scale, in private or in public. Given the transnational nature of the environmental problems facing humanity, it seems reasonable to adopt a notion of citizenship which extends possibilities for participative action to all people in all areas of life. It is this conception of citizenship which Dobson (2003) calls 'ecological citizenship', and it represents a clear departure from Sagoff's dualistic understanding of private and public interests and activity: ecological citizenship explicitly defines private 'consumer' behaviour as political and a space for collective action for the common good. In this way, ecological citizenship rises above traditional understandings of citizenship to embrace new possibilities, in particular the development of consumption as a site of political activity and sustainable consumers as a key element of government strategy. What then are the obligations of an ecological citizen?

Dobson's ecological citizenship uses the 'ecological footprint' metaphor (Wackernagel and Rees, 1996) as a touchstone for understanding the obligations of ecological citizens as a justice-based account of how we should live, based upon private and public action to reduce the environmental impacts of our everyday lives on others. In this model, each of us is responsible for taking up a certain amount of ecological 'space' in the sense of resource use and carrying capacity burden, and this space is expressed as a footprint on the Earth. It is assumed that there is a limited amount of ecological space available, which when equitably distributed among all inhabitants delivers an allocation of 1.8 global hectare per person. However, the footprint of the average global citizen is 2.2 gha, indicating an ecological overshoot, but this is not distributed evenly: UK residents require 5.6 gha, and in the USA it is 9.6 gha (compared to 0.8 gha in India) – therefore current distributions are unjust and inequitable (WWF, 2006). The ecological footprint of a western consumer includes areas spread across the globe, and impacts upon people distant in space and time. The footprints of people within industrialised nations are much larger than that of, and indeed have negative impacts upon the life chances of, the inhabitants of developing countries.

In this way, environmental and social inequity and injustice is visualised. An ecological citizen's duties are therefore to minimise the size and unsustainable impacts of one's ecological footprint – though what is 'sustainable' is of course a normative rather than technical question (Dobson, 2003). Ecological citizenship is non-territorial and noncontractual and is concerned with responsibilities and the implications of our actions on the environment and on other, distant people; a similar model, called 'planetary citizenship' is put forward by Henderson and Ikeda (2004). The challenge is to find mechanisms and initiatives and a meaningful social context both for developing ecological citizenship, and for expressing ecological citizenship in daily life – in the supermarket, the classroom, the household and the workplace – in other words, to enable and encourage people to act as ecological citizens and reduce their ecological footprints and specifically to overcome the limitations of the mainstream sustainable consumption strategy outlined above (Dobson, 2003; Seyfang, 2005, 2006a). The New Economics aims to meet that challenge, through the development of new institutions and socio-technical infrastructure that allow the expression of ecological citizenship.

Making it count: evaluating sustainable consumption

The New Economics presents many challenges to mainstream thought and practice on sustainable consumption; its objectives are to develop a practical approach to sustainable development which encompasses new definitions of wealth and work, new uses of money and which integrates ethics into economic life, and thereby to provide ecological citizens with the means to express their values and reduce their ecological footprints. It champions the active ecological citizen rather than the green consumer, placing at centre stage actors 'able to vote with more than their feet in support of collective projects like those of environmental reform, [and] who have a hand in shaping options as well as exercising choice between them' (Shove, 2004: 115, 116). The New Economics aims to deliver this through shifting the trajectory of economic development towards an alternative goal of sustainable well-being, and enabling the emergence of a range of new infrastructures of provision based on these values. Its theories of behaviour change are pluralistic: its strategies incorporate simple cognitive behaviour-change incentives, as well as responses to more complex social and psychological contexts, but throughout this body of thought is a fundamental recognition that existing infrastructures of provision are not fit for purpose. As such, it aims to trigger, enable and support a series of socio-technical transitions in mainstream regimes, each of which is comprised of interrelated technologies, institutions, norms, cultures and expectations. This process is discussed in depth in the next chapter, but here the task is to consider what such changes might look like in practice: in other words, how are we to evaluate New Economics initiatives for sustainable consumption?

Despite a growing number of practical applications of this model, there is a paucity of robust empirical research to test the ideas of this New Economics approach, and there has to date been no systematic means of evaluating activities to assess their contribution to sustainable consumption. To meet that need, therefore, here we present a new qualitative evaluation framework which is designed to incorporate the key elements of the New Economics vision of sustainable consumption, and builds on the theoretical foundations outlined above. A New Economics strategy for sustainable consumption would therefore embody the following five characteristics: localisation, reducing ecological footprints, community-building, collective action, and building new infrastructures of provision. This set of indicators (outlined in Table 3.1) forms the basis of a multi-criteria evaluation tool for sustainable consumption, which is applied to the initiatives examined in the chapters to follow.

Table 3.1   Indicators of sustainable consumption

Indicator	Description	Example
Localisation	Making progress towards more self-reliant local economies; import-substitution; increasing the local economic multiplier; reducing the length of supply chains.	Supporting local businesses; eating more local, seasonal food to cut food miles; encouraging money to circulate locally; 'buy-local' campaigns; DIY, growing food on allotments.
Reducing ecological footprints	Shifting consumption to cut its social and environmental impact on others, to reduce the inequity of current consumption patterns; cutting resource use; demand-reduction; carbon-reduction and low-carbon lifestyles.	Downshifting; voluntary simplicity (accepting cuts in income in return for higher quality of life and lower consumption); energy and other resource conservation e.g. water-saving devices, energy efficiency and insulation, buying local to reduce transport costs; choosing ethical and fair trade where possible; sharing goods instead of owning them; cutting consumption; choosing less carbon-intensive goods and services; avoiding flying.
Community-building	Nurturing inclusive, cohesive communities where everyone's skills and work are valued; growing networks of support and social capital; encouraging participation to share experience and ideas.	Developing social networks around green building, local food, community volunteering; overcoming social exclusion barriers to participation; fostering shared experiences through group activities; growing friendships.
Collective action	Enabling people to collaborate and make effective decisions about things which affect their lives; changing wider social contexts by institutionalisation of new norms; active citizenship.	Boosting self-efficacy and empowerment; encouraging participation in local organisations; engaging with local government and public policy; generating critical mass so that new sustainable behaviours become the norm.
Building new infrastructures of provision	Establishing new institutions and socio-technical infrastructure on the basis of New Economics values of wealth, work, progress and ecological citizenship.	Alternative food systems which avoid supermarkets; autonomous housing which doesn't rely on mains services; new systems of exchange which value abundance and reward sustainable consumption.

4
Grassroots Innovations for Sustainable Consumption

(with Adrian Smith)

Everybody, it appears, is committed to sustainable consumption; but not everybody is seeking it in the same way. Moves towards sustainability are generating a variety of social innovations as well as innovative technologies – new organisational arrangements and new tools – in different arenas and at different scales to address consumption issues. Grassroots, niche innovations of the type discussed in Chapter 3 differ from commercial business reforms such as those favoured by mainstream policy in Chapter 2; they practise quite different kinds of sustainable consumption. There is a qualitative difference between, for instance, a community-supported organic vegetable box scheme and the range of organic products sold at a supermarket; the social, economic and environmental dimensions of sustainable consumption are traded-off differently. In order to better understand the role and potential of community-based New Economics efforts to move towards sustainable consumption patterns, this chapter examines the characteristics of grassroots initiatives, and sets out a new conceptual model which views them as specifically *innovative* activities.

Grassroots action for sustainable consumption takes different forms, from furniture-recycling social enterprises to organic gardening cooperatives, low-impact housing developments, farmers' markets and community composting schemes. While community action addresses local problems, these are not irrelevant to wider contexts: 'the global problems or perspectives are ... "translated" and fitted into the local, specific circumstances of the individuals' (Georg, 1999: 460), for example through efforts to reduce personal carbon emissions to address climate change. Since 1992, over 400 local authorities in the UK produced Local Agenda 21 strategies, alongside growth of independent, community-based work on 'local sustainability'; Shell Better Britain's network of groups, for instance, grew from 10,000 in 1992 to 26,000 in 2002 (Church and Elster, 2002). Rarely has the innovativeness of this activity been acknowledged. The term 'grassroots innovations' is used here to describe networks of activists and organisations generating novel bottom-up solutions for sustainable development and sustainable consumption; solutions that respond to the local situation and the interests and values of the communities involved. In contrast to mainstream business greening, grassroots initiatives operate in civil society arenas and involve committed activists experimenting with social innovations as well as using greener technologies.

Reflecting this disparity are two parallel policy strands within the UK's sustainable development strategy[1] (HM Government, 2005). These strands are: a) ecological modernisation and technological innovation, and b) community action and the social economy; each strand has traditionally been studied in separate literatures. This chapter makes the case that this division inhibits understanding of the innovative potential of grassroots initiatives, and prevents us appreciating its full potential for change. It bridges that divide and integrates these two previously unrelated areas, in order to offer an original theoretical approach to the analysis of community-level action for sustainable consumption. This new agenda considers the grassroots a neglected *site of innovation* for sustainability, hitherto eclipsed by green reforms in more conventional business settings. Viewing community-level activities as *innovative niches*, affords a better understanding of the potential and needs of grassroots initiatives, as well as insights into the challenges they face and their possible solutions, and the scope for scaling up and diffusing these innovations into wider society. This integrated analysis provides a foundation for exploring policy, theory and practice in the chapters to follow, and for asking questions about how small-scale grassroots efforts for sustainable consumption might grow and influence the mainstream.

---

[1]Whilst this chapter focuses upon the UK, it is of wider relevance. The Plan of Implementation of the World Summit on Sustainable Development in 2002 contains both innovation policy and community action commitments.

Sustainable development contexts: innovation and community action

The language of innovation is common in sustainable development policy contexts. For instance, the UK strategy for sustainable development 'Securing the Future' states 'The goal of sustainable development is to enable all people throughout the world to satisfy their basic needs and enjoy a better quality of life without compromising the quality of life of future generations', and this will be pursued 'through a sustainable, innovative and productive economy' (HM Government, 2005: 16). The government pursues an 'ecological modernisation' agenda (Murphy, 2000) through its strategy for Sustainable Production and Consumption, seeking 'greener' markets using taxes, incentives and better information, and so encouraging technological innovation to improve resource efficiency and decouple economic growth from environmental degradation (DEFRA, 2003b). Innovation is defined as 'the successful exploitation of new ideas – incorporating new technologies, design and best practice [which] is the key business process that enables UK businesses to compete effectively in the global environment' (DTI, 2005). Government makes the link with sustainability in the 2003 Innovation Report, stating innovation will be essential for meeting the environmental challenge (DTI, 2003a). In this vein, 'sustainable innovation', 'eco-preneurship', and eco-efficiency are key terms used to describe greener business activity (Beveridge and Guy, 2005; Fussler and James, 1996), and espoused by bodies such as the World Business Council for Sustainable Development (Holliday and Pepper, 2001: 3). Alongside greener business innovation, the government aims to promote sustainable consumption through 'market transformation', and the development of more sustainable market choices for products and services (DEFRA, 2003b).

However, the UK strategy also recognises the contribution made by small-scale local activities, and has a particular emphasis on *delivery* of sustainable development at all scales. Prime Minister Blair stated: 'Many local communities understand the links between the need to tackle national and global environmental challenges and everyday actions to improve our neighbourhoods and create better places to live... I want to reinvigorate community action for sustainable development' (HM Government, 2005: 29). A new initiative,

'Community Action 2020', will build on Local Agenda 21 to be 'a catalyst for thinking globally and acting locally in communities across England' (*ibid.*). It promotes local food initiatives, community energy efficiency schemes, recycling projects and Fairtrade activities, plus participation in decision-making, volunteering, capacity-building, information-sharing and community mentoring (DEFRA, 2005b). This represents a growing policy focus on the *social economy* (the 'third sector' between the private and public sectors, comprising social enterprise, cooperatives and mutuals plus voluntary and community organisations) as a source of sustainability transformation, active citizenship, and public service delivery (HM Treasury, 2002a). The social economy represents 'a wide family of initiatives and organ- isational forms – i.e. a hybridisation of market, non-market (redistri- bution) and non-monetary (reciprocity) economies' (Moulaert and Ailenei, 2005: 2044; see also Dobson, 1993). As such it covers a diverse range of activities, from charities and cooperatives which may operate as commercial businesses, through to recycling networks such as Free- cycle, and Local Exchange Trading Schemes. This amounts to a sig- nificant level of activity, as a range of studies indicate. Kendall and Almond (1999) calculate that the UK civil society sector employs the equivalent of 1.4 million full-time employees (5% of the economically active population) and benefits from the unpaid efforts of the equiva- lent of 1.7 million full-time volunteers (5.6% of the economically active population), and contributes 6.8% of GDP.

Focusing on the social economy, the UK strategy specifically high- lights community engagement in governance as a key element of a sustainable society (HM Government, 2005; Seyfang, 2006b), and looks to community and voluntary groups to lead the way and gen- erate the innovations in governance, behaviour and lifestyle changes – embedded and 'owned' in local communities – necessary for sus- tainable consumption and production (DEFRA, 2005b; Rogers and Robinson, 2004). In addition, DEFRA is developing its own strategy to support social enterprise because of the ways the sector combines social, economic and increasingly environmental objectives, and con- tributes directly to its strategic goals of achieving sustainable rural communities, waste reduction, biodiversity enhancement, action on climate change, and so on (DEFRA, 2005a). This policy strand further- more considers social structures by acknowledging that 'We need to understand more about the social and cultural influences which shape our consumption choices, habits and impacts' (HM Government, 2005: 51–2). Whilst Community Action 2020 lists actions which reshape social infrastructures of provision (DEFRA, 2005b), policy also acknowledges the role of 'socio-technical regimes' which influence behaviour, constrain individual choice sets and limit the transformative potential of the market (Maniates, 2002; Jackson and Michaelis, 2003).

Community action is becoming embedded in sustainability policy for a variety of reasons (DEFRA, 2005a). Principal among these is the need for active citizens and strong local democratic institutions to 'own' and embody sustainable development (Young, 1997). Locally-rooted action generates socially-embedded changes in behaviour (Burgess *et al.*, 2003). Overlapping these are government objectives to boost social capital through micro- and meso-level activities (PIU, 2002) and the emerging policy agenda for decentralisation and the 'New Localism'. To this one could also add the (problematic) policy agenda for 'outsourcing' traditional welfare state functions to community groups.

In sustainable innovation policy, recent statements and initiatives seeking to open developments to wider stakeholder participation are found, including citizens and local communities. Policymakers acknowledge 'increasing aspirations towards public accountability and democratic control of the direction of development of science and technology' (DEFRA, 2004: 16). Public engagement is on the agenda, rhetorically at least (Stirling, 2004; Wilsdon and Willis, 2004), and threads within UK innovation policy are converging in a way that provides potential opportunities for grassroots innovation, and distinguishes this from an earlier generation of citizen science and alternative technology (Irwin *et al.*, 1994; Corborn, 2005; Winner, 1979; Boyle and Harper, 1976; Smith, 2004).

Understanding sustainable innovations

Attention now turns from the policy context to ideas in the sustainable innovation literature. Radical improvements in production and consumption *systems* (e.g. 'factor 20' resource efficiency or 60% carbon emissions reductions) imply greener innovation different from traditional improvements to single products or business practices; innovation is needed at the scale of 'socio-technical regimes' (Berkhout,

2002). Transforming systems of production and consumption poses considerable challenges; innovation studies identify mutually reinforcing processes that tend to channel developments along trajectories (Nelson and Winter, 1982; Dosi *et al.*, 1988; Russell and Williams, 2002). Changes tend to be incremental and path dependent owing to:

- the cognitive frameworks, routines, resources, capabilities, and knowledge of technology producers and users, and expectations about what kinds of knowledge will be profitable in the future (Nelson and Winter, 1982; Dosi, 1982);
- the way specific social and technical practices are embedded within wider, facilitating infrastructures, which subsequently restrict opportunities for alternatives (Jacobsson and Johnson, 2000);
- incumbent practices enjoy economies of scale (e.g., mass markets) and positive network externalities (it is easier and less risky to follow established practices than to invest in new practices) (Arthur, 1988; Dosi, 1982);
- the co-evolution of institutions with technological practices, like professional associations, government policies, and market rules reinforce existing trajectories (Hughes, 1983; Walker, 2000);
- prevailing market and social norms influence the kinds of performance deemed satisfactory, and the lifestyle routines and norms that develop embed these practices further (Yearley, 1988; Shove, 2003).

In short, entrenched cognitive, social, economic, institutional and technological processes lock us into trajectories and lock out sustainable alternatives. The term 'socio-technical regime' captures this complex configuration of artefacts, institutions, and agents reproducing technological practices. The socio-technical 'adjective is used to stress the pervasive technological mediation of social relations, the inherently social nature of all technological entities, and indeed the arbitrary and misleading nature of distinctions between "social" and "technical" elements, institutions or spheres of activity' (Russell and Williams, 2002: 128). The development of the socio-technical is a highly social, collective process, and ultimately it is diverse social actors who negotiate innovation (Smith *et al.*, 2005). Imposing a normative goal like sustainable development upon existing socio-technical regimes implies connecting and synchronising changes amongst actors, institutions and artefacts at many different points within and beyond the regime.

Consider the co-housing model. It is a model of community structure whereby residents live in houses around a 'common house'. This common house contains a large kitchen and dining area for shared meals, and industrial-sized washing machines and lawnmowers. Cars are kept to the perimeter (and may be shared), allowing for open gardens and footpaths between houses. This structure combines privacy with communal activities (planning meetings, weekly shared meals, easy conviviality, supportive networks of neighbours), and potentially reduces overall consumption. It is essentially a social innovation – a restructuring of the social institutions of housing – rather than a technological one (Hines, 2005; Meltzer, 2005). However, it opens up terrain for more sustainable technologies. Co-housers can pool resources for the use of small-scale renewable energy technologies, rainwater harvesting, grey water recycling, and more sustainable construction materials and designs unavailable to individual households. In short, social innovations and the diffusion of technological innovations are intimately linked.

Historically regimes do undergo radical change. Succession tends to begin within a network of pioneering organisations, technologies and users that form a *niche* practice on the margins. Niche situations (e.g., unusual applications, demonstration programmes, social movements) provide space for new ideas, artefacts, and practices to develop without full exposure to the range of processes channelling regime development (Schot, 1998; Geels, 2004; Rip and Kemp, 1998). Hoogma *et al.* (2002: 4) state: 'A niche can be defined as a discrete application domain ... where actors are prepared to work with specific functionalities, accept such teething problems as higher costs, and are willing to invest in improvements of new technology and the development of new markets.' If successful, alternatives become sufficiently robust to develop niche markets, branch out, and attract mainstream interest (Schot *et al.*, 1994). This perspective informs certain approaches to sustainable development which are based upon the strategic creation of green niches that inform possibilities for more sustainable regimes (Kemp *et al.*, 1998; Smith, 2004). Green niches are sustainability experiments in society in which participation is widespread[2] and the focus is on social learning. Niche-based approaches explore problem framings (e.g., mobility, food, energy services) and search for solutions – in contrast to technology demonstration projects that begin with 'technical solutions' to tightly framed problems. Niche practices that resonate with widespread public concern sometimes catch on, get copied, become adapted and spread.

Niche-based advocates qualify their bottom-up enthusiasm. Niches alone will not seed wider change (Hoogma *et al.*, 2002). Work on multi-level socio-technical change identifies tensions and contradictions within incumbent regimes, exacerbated by pressures deriving from broader socio-economic dynamics, as opening niche opportunities and driving the transformations (Geels, 2004). Social movement agitations against regimes contribute to these pressures, but are distinct activities from the grassroots innovations considered here. Oil shocks, demographic change, economic recessions and so on are more general sources of pressure or shock on regimes. Change depends upon contingencies and processes beyond the unilateral control of niche actors (Berkhout *et al.*, 2004). Niches still play a role as sites where alternatives try to resolve regime contradictions. Niches are potential sources of innovative ideas, even if not models or blueprints (Smith, 2006). More pragmatic, intermediary initiatives involving the mainstream help spread ideas and practices, but involve compromises and mutual adjustments that nevertheless take important cues from green niches. Ecopreneurs and intermediary organisations more attuned to market and commercial imperatives assist this bridging activity.

For example, East Anglia Food Link (EAFL), a small sustainable food NGO, began promoting locally sourced organic food in schools and hospitals in 1999. Marginal successes accrued over the following years, but in 2005 the national agenda on public sector catering was rewritten after a high-profile TV series criticising the standard of food in schools. This galvanised public opinion and spurred government policy changes that encouraged local, freshly made organic food. EAFL, along

---

[2]Kemp *et al.* (1998: 188) argue the niche-based approach is the 'collective endeavour' of 'state policy-makers, a regulatory agency, local authorities (e.g., a development agency), non-governmental organizations, a citizen group, a private company, an industry organization, a special interest group or an independent individual'.

with other Food Link organisations, were identified as pioneering sources of good practice (Wakeman, 2005). EAFL's approach is a radical departure from mainstream food and farming policy, reflecting quite different values, beliefs about the environment, and desirable sustainability outcomes (Seyfang, 2007). An organic farmer cooperative supplying local markets and delivering direct to households, schools and hospitals is experimenting not only with food production techniques, but with the social infrastructure of food supply. It offers a hitherto absent alternative to mainstream food, one which responds to the logic of internalising the environmental and social costs associated with globalised food systems (Pretty, 2002; Seyfang, 2006a; see chapter 5).

As an analytical framework, the niche-based approach studies niche emergence and development (Smith, 2007). Analysis focuses upon the social networks, learning processes, expectations and enrolment of actors and resources in emerging niche practices. Armed accordingly, advocates recommend policies to improve the development and influence of niches, including nurturing diverse niches, facilitating greater actor interaction, promoting social learning, and seeking institutional changes that embed promising lessons (Kemp *et al.*, 1998; Smith, 2007; Hoogma *et al.*, 2002). Lessons derived from the niche need not be restricted to narrow, technical appraisals of performance. Such 'first-order' learning can be supplemented by 'second-order' learning that generates lessons about the alternative socio-cultural values underpinning the niche and implications for diffusion (Hoogma *et al.*, 2002). Insights into deeper institutional changes can be complemented by lessons relating to the constituencies, capabilities, contexts and markets able to appropriate niche elements (Weber *et al.*, 1999; Kemp *et al.*, 1998; Hoogma *et al.*, 2002). As such, niche-based approaches demand an interactive policy style mature enough to recognise the value in acknowledging and learning from failure as well as success. Elements of niche practice that do not 'work' can be just as informative for sustainable developments as those aspects that operate successfully.

Contrasts between green niches and mainstream regimes can already be drawn in many systems of production and consumption, such as housing, food, energy and banking. This niche-based analytic and policy perspective might also encourage fresh thinking about grassroots initiatives. Can the grassroots be conceptualised as a site for innovative niches? Whilst the literature on green niches did not develop with an explicit focus on grassroots innovation in mind, early case studies included grassroots initiatives (e.g., wind energy in Denmark, car clubs in Switzerland) (Kemp *et al.*, 1998; Hoogma *et al.*, 2002).

Some characteristics of grassroots innovations

The niche framework provides a potentially fruitful bridge between analyses of grassroots initiatives as civil society activities and a role for them in sustainable innovation policy. Here the conceptual model of green niches is extended to the grassroots realm, from the market economy to the social economy, with sensitivity to the fundamental differences between the two sectors. It is important to qualify this potential; grassroots innovations are not the exclusive, powerful vanguard for more sustainable futures, but a source of innovative diversity.

Sustainable innovation traditionally deals with niches within the market economy. Sustainable innovation is sheltered from the full extent of market competition through a system of tax breaks and subsidies, to allow development until it can compete in the market. Niches are spaces where 'the rules are different', and conventionally these rules are those of the market. Grassroots innovations, in contrast, exist within the social economy of community activities and social enterprise. The social economy differs from the market economy; appropriation of profits by capital under the latter is suspended in favour of reinvesting any surplus into the grassroots under the former (Amin *et al.*, 1992). Relevant to our niche perspective is the way grassroots initiatives also emphasise different social, ethical and cultural rules. For example, community currencies are new forms of money designed to serve social, economic or environmental purposes which conventional money does not, and so reward specific types of behaviour. The NU Spaarpas green loyalty card piloted in the Netherlands awards points for purchasing local, organic or fair trade products, or for recycling household waste; the points are redeemed for public transport tickets, or discounts off green services. In this way, it sets up incentives different from the mainstream economy (Seyfang, 2006c; see chapter 7).

The institutional form of conventional innovations appears straightforward; firms generate financial income commercially, from selling the products they innovate. The driving force is profit; firms seek to appropriate the benefits of innovation in order to move ahead of the competition and so capture market rents (Schumpeter, 1961). Competitors innovate too, and rents are gradually eroded, inducing further innovation. Obviously, there are complexities and nuances associated with this basic logic,[3] but by situating itself within conventional market economics, the sustainable innovation literature has to align with it. Green market-based niches will, ultimately, only prosper if they can attract significant investment and business commitments, which will only happen if the niche innovation can demonstrate a highly profitable potential compared to other (unsustainable) opportunities for capital.

The institutional forms for grassroots innovative niches are also complex, but in different ways. There are diverse organisational forms: cooperatives, voluntary associations, mutuals, informal community groups, social enterprises. Their resource base is similarly pluralistic, including grant funding, limited commercial activity, voluntary input and mutual exchanges. The spectrum of organisations exhibit varying degrees of professionalisation, funding and official recognition. Chanan (2004) finds four out of five identifiable groups in the grassroots sector are small, low-profile, voluntary, citizen-led and community-driven groups (*cf.* high-profile professionally-led voluntary organisations). Official and quasi-official groups operate alongside informal, voluntary activities, and their relationships can be both complementary and competitive. Grassroots innovations are driven by two motives more forgiving towards sustainable innovation compared to rent seeking firms. These are social need and ideology. Meeting social (and environmental) needs is the primary function. The social economy provides flexible, localised services in situations where the market cannot. Incumbent production and consumption systems fail some communities, perhaps because groups are socially and economically disadvantaged, unable to access goods, services and markets, or because market choices do not extend to sustainability, such as fresh, local organic food in season, or autonomous housing, or community renewable energy (Maniates, 2002; Manno, 2002).

---

[3]In practice, market power can prevent perfect competition. The ability to 'catch-up' depends upon resources, institutions, and abilities to appropriate benefits (Clark, 1985).

However, niche approaches must not condemn to the margins people who do not wish to be there; grassroots participants might actually dream of mainstream consumption, but for reasons of social and economic exclusion, find themselves in a niche instead, e.g., furniture recycling. Many initiatives in excluded communities seek to build capacities for entering the mainstream. Local Exchange Trading Schemes (LETS), a type of complementary currency, have been advocated as a tool to build the skills, confidence and social contacts for people to enter the formal employment market (Williams *et al.*, 2001; Seyfang, 2001c; see chapter 7).

Unmet social need is not the sole grassroots driver; ideological commitment to alternative ways of doing things is another. Such ideologies run counter to the hegemony of the regime, and some grassroots innovations develop practices based on reordered priorities and alternative values. As this book demonstrates, 'New Economics', for example, proposes a socio-economic system geared towards quality of life rather than economic growth *per se*, and favours localised, self-reliant economies as the basis of sustainable communities (Jackson, 2004a; Robertson, 1999). This can be expressed through init-

Table 4.1   Comparing the characteristics of market-based and grassroots innovations

	Market-based Innovations	Grassroots Innovations
Context	Market economy	Social economy
Driving force	Profit: Schumpeterian rent	Social need; Ideological
Niche	Market rules are different: tax and subsidies temporarily shelter novelty from full forces of the market	Values are different: alternative social and cultural expressions enabled within niche
Organisational form	Firms	Diverse range of organisational types: voluntary associations, coops, informal community groups
Resource base	Income from commercial activity	Grant funding, voluntary input, mutual exchanges, limited commercial activity

iatives like locally-produced food, or by rewarding socially reproductive labour not valued in the formal labour market (Seyfang, 2006a, b). Niches can emerge in explicit opposition to mainstream regimes. The organic movement began with idealists committed to healthy, local food economies in opposition to the industrialisation of food.

In summary, key comparisons between niche innovations in the market economy and the social economy are shown in Table 4.1.

Grassroots innovative potential

The theory on niches discussed above identifies two types of benefit: *intrinsic* benefits; and *diffusion* benefits. They are not mutually exclusive, and overlap in practice. However the distinction is useful conceptually. One values the niche for its own sake (intrinsic benefits), the other as a means to an end (diffusion benefits). The distinction delineates 'simple niches' (not seeking regime change) from 'strategic niches' (seeds for wider transformation).

Intrinsic benefits

The principal intrinsic benefit relates to the social and environmental basis of the niche. But what can small-scale community action contribute to sustainable development? A review of grassroots action for sustainability by Church and Elster (2002) identified a range of direct environmental benefits such as reduced car-use, increased recycling, and planting trees. When assessing impacts, they note 'small local projects may seem almost irrelevant at city-scale or above, but if wider policies lead to larger numbers of them, there is every reason to expect them, in aggregate form, to have proportionate impact' (Church and Elster [2002: 25], citing the Community Recycling Network comprising 350 local initiatives). They also identified significant socio-economic impacts with benefits for sustainable communities. These related to job creation, training and skills development, personal growth (e.g., self-esteem and confidence), a sense of community, social capital, improved access to services and facilities, health improvements, and greater civic engagement. Integrating small-scale renewables into community projects brings similar benefits (Devine-Wright, 2006).

The self-image of these initiatives is not as environmental organisations, but rather as groups aiming to improve quality of life in local communities. This is an important point. Grassroots initiatives need not consciously practice 'strong' sustainability for them to have an impact concordant with those objectives. Groups doing 'simple' activities like furniture recycling, community composting, or running a volunteering project, may nevertheless develop significant sustainability practices. Of course, sustainability is a contested concept, and diverse 'sustainabilities' are being experimented with at the grassroots and in other domains. Some practices run counter to certain forms of sustainability; consider the way extreme localism/ autonomy projects conflict with sustainable developments conceived for poorer regions through Fair Trade. The point is to appreciate empirically the sustainability dimensions and trade-offs being developed in niches, and to relate niche self-interpretations of performance to their motivating ideologies.

Grassroots innovation can deliver sustainability benefits where top-down measures struggle. This is because community action utilises contextualised knowledge and implies a better 'fit' of solution (*cf.* inflexible top-down targets and procedures) (Burgess *et al.*, 2003). Grassroots groups have experience and knowledge about what works in their localities, and what matters to local people. They can be well-placed to present sustainability issues in ways more meaningful, personal and directly relevant, and which 'goes with the grain of people's lives' (Roberts, 2005). They can engage and reinforce behavioural change.

The grassroots can also be a site for action on 'unpopular' or 'fringe' issues not taken up by mainstream actors. A 'world within a world', grassroots innovations are a demonstration that another way is possible, building alternative infrastructures to the existing regime. However unlikely mainstream diffusion, the niche nevertheless stands as a symbolic embodiment of alternatives (Amin *et al.*, 2002; Leyshon *et al.*, 2003). Wakeman (2005) uses the metaphor of a 'green conveyor belt' to express the notion that while some grassroots innovations begin in niches, then grow and are incorporated into mainstream regimes (such as organic food), radical action on unfunded issues continuously regenerates at the grassroots.

Diffusion benefits

In alternative green niches, people's motivations for action are based upon different values from the mainstream. This represents the bottom-up generation of alternative systems of provision, vertical commodity chains (comprising production, marketing, distribution, retail and consumption in social and cultural context) which mediate between and link 'a particular pattern of production with a particular pattern of consumption' (Fine and Leopold, 1993: 4). For example, Time Banks are community-building projects where participants give and receive services in exchange for time credits. Everyone's time is valued equally, and taken-for-granted (but sometimes scarce) skills and abilities, such as time for listening sympathetically, companionship, doing someone's shopping, walking a dog, light gardening or home repairs, are recognised, valued and rewarded. The values expressed through this time-based system of exchange contrast with the conventional economy; they value all productive labour equally (Boyle, 2005). So while participants enjoy the social networking, sense of being useful, and opportunity to help others, they are also imbued with alternative values relating to the nature of work, how people are valued as assets; they respond to incentives to perform the types of neighbourhood work needed to build healthy communities. The alternative metrics expressed in this Time Banks niche are expanding as a network of small-scale projects that demonstrate how measuring 'wealth' and 'sustainability' is a matter of perspective. Indeed, the UK government's sustainable development strategy calls for new research to define 'wellbeing' in place of economic growth (HM Government, 2005).

In such cases, grassroots activists seek to mobilise communities to create new 'systems of provision'. These grassroots innovations offer the potential to generate transformations in production-consumption systems in a way that individuals cannot (Maniates, 2002). By joining small, everyday decisions about food, say, for whatever reason (taste, health concerns, food miles, supporting local growers), communities of citizens participate in that (radical) creative process (Dobson, 2003). As such, they represent collective efforts to transform not simply the market choices available, but sometimes the entire market system itself. They help overcome the principal problem with an individualised approach to greening the market, namely that acting individually, consumers are powerless to change the rules of the game, they are stuck within current socio-technological regimes (Seyfang, 2005, 2006a,c). Grassroots innovations can have ambitions beyond the micro level. Some seek new institutions based upon different values from the incumbent regime, and hence contribute critically towards change at the regime level too.

Perceived as niche initiatives in an alternative kind of sustainable development (*cf.* mainstream business reforms), grassroots innovations might also hold some comparative power. By looking at the kind of practical sustainability expressed in these niche initiatives, more mainstream green reformers, and their critics, might obtain a different perspective upon mainstream efforts. Somewhat analogous to travelling through another country and culture, the experience causes us to reflect upon our home culture. The niche model might prove effective precisely because it draws contrasts. It could serve as a dialogical device for reflecting critically upon mainstream reforms. Stark contrasts between niche and mainstream, whilst making the translation of lessons from niche to mainstream difficult (see below), can still provide a basis for critical reflection.

In niche terms, grassroots initiatives exhibit first- and second-order learning. They build environmental support and capacity. Practices develop that provide services with reduced environmental impact whilst, at the same time, encouraging participants to further reflect upon how their need for services is framed and developed in other areas. Church and Elster (2002) identify a wide set of indirect environmental and social impacts from grassroots innovations, for example environmental awareness-raising, education and promotion, changing the attitudes of local policymakers, engaging people in sustainability issues in their daily lives, and developing new ways of working towards sustainable development. As a result of niche practices, which are often participative, individuals and communities can benefit in terms of greater empowerment and confidence, skills and capacity for further community-based action.

Challenges faced by grassroots innovations

Whilst grassroots innovations hold normative promise, they are not a panacea. It is important to analyse their problematic challenges, which can be similarly categorised as *intrinsic* and *diffusion-related*.

Intrinsic challenges

Challenges confront grassroots innovations from their inception; establishing an initiative requires a particular combination of skills, key individuals and champions, resources and supportive contextual factors. After start-up, the challenge is to survive and keep going, which requires additional skills and people, plus resilience and a resource base. Dilemmas arise over whether to try to commercialise (presenting diffusion challenges, see below) or to engage with government support programmes. Grant funding and voluntary activity, common amongst grassroots innovations, pose significant problems. Funding programmes are often short term, frequently linked to constraining targets, bureaucracy and requirements, and leave little room for core development (support programmes for community renewables being a prime example). Frameworks for funding are often imposed by funders, rather than responding to recipients' development. Grassroots innovations can fall between the interstices of traditional social, economic and environmental issue boundaries. Their 'institutional fit' with departmental-based funding regimes can be poor, resulting in difficulty combining and fulfilling the distinct criteria of multiple, single-issue funders.

Experience suggests initiatives spend 90% of their time simply surviving, and only 10% developing the activity (Church, 2005; Wakeman, 2005). This has implications for niche survival. First, they fail to develop robustness and resilience to shocks like funding cuts, key people leaving, turnover of volunteers, burn-out of activists, shifts in government policy. Secondly, short-lived initiatives frequently leave no formally documented institutional learning. The skills and learning are tacitly held within people, rather than being consolidated in readily accessible forms.

Niches at the grassroots level are interdependent upon technology developers, and provide sites where emerging sustainable technologies find application and development. Yet grassroots innovators, like others, are technology takers initially, and can struggle to identify and obtain appropriate sustainable technologies. This interdependency could be made more effective by opening participation in technology development to grassroots innovation. The challenge is considerable, especially where technology development is transnational. Appliance recycling initiatives, for example, reveal considerable insights into design for repairability and remanufacture, but this needs conveying to the product development decision-makers of manufacturers whose headquarters may be in a different country.

Diffusion challenges

Grassroots influence is limited by a number of factors. First, small-scale and geographical rootedness makes scaling up difficult. Niches need reinterpreting and transposing for other scales. Whilst policy interventions can bridge niche and mainstream situations, they can also filter and reformulate the practices that work on wider scales. Alternatively, small-scale initiatives can reproduce elsewhere by ensuring groups are well connected regionally and nationally. For instance, Time Banks operate successfully at a small scale, allowing participants to feel that they know most of the other members; they grow by 'budding off', to retain the sense of neighbourliness, and keep coordination manageable (Boyle, 2005).

Paradoxically, a key benefit of grassroots innovations, namely the 'world within a world', undermines diffusion. Whilst practices where 'the rules are different' have certain strengths, those strengths become barriers when in concerted opposition to incumbent regimes. In these instances, there is an important distinction between communities of location (geographically-based grassroots groups meeting a social need) and communities of interest (ideologically-based initiatives). One cannot assume that grassroots innovations and local action is always socially cohesive. Ideological niches define themselves as 'other' or 'alternative' to the mainstream – an identification that makes outreach and diffusion difficult. This contrasts with the niche literature, which argues that successful influence requires a degree of congruence with regime practices if niches are to have a chance of catching on (Hoogma *et al.*, 2002; Weber *et al.*, 1999). A corollary is that compatibility limits the degree to which green niches can diverge radically from the mainstream, thus blunting their radical potential (Smith, 2006).

However, even radical green niches can eventually exert influence upon the mainstream, though not in forms anticipated by original niche idealists. *Elements* of niche practice that can be adapted and accommodated easily within the market are appropriated when the regime feels pressure for sustainable reforms. In this way, grassroots initiatives remain sources of learning, even if it is only the more appropriable, marketable lessons that spread. The form of sustainability that diffuses alters (reduces) accordingly. The inability of the more complete versions of radical sustainability to diffuse from the niche suggests both the limited power of the niche and limited capacity of the incumbent regime to become more sustainable. Conflict arises between those wishing to remain 'purist' and others seeking wider yet partial influence (system-builders) and prepared to compromise. System-builders might be welcomed as recognition of the worth of the niche, but also resented as an unwelcome sellout to economic interests. Niche pioneers can be pushed aside by the entry of more powerful commercial interests practicing a more limited proxy to niche activities (but which reaches further, e.g., large waste management companies developing kerb-side recycling activities to the detriment of earlier, less capitalized community-based operations).

A further challenge is policymakers' risk aversion. Innovation is an experimental process, and an important aspect of this is openness to learning from failure. The policy culture is insufficiently mature to identify this as a positive process. Funding constraints inhibit experimentation and punish failure by withdrawal of resources. The challenge is to develop support mechanisms that allow grassroots initiatives to revise and continue in the light of earlier difficulties, and diffuse the lessons learnt. Whilst continued funding of failure can be difficult to justify, it seems unreasonable to cut funding from initiatives willing to adapt activities, overcome earlier problems, and continue experimenting. This is the lifeblood of inno-vation.

Finally, there is a wider, institutional challenge. Change at higher levels – within incumbent regimes and overarching socio-economic processes – opens opportunities for niche diffusion. Sustainability pressures can spur regime actors into appropriating greener activities from niches. Church (2005) argues that local action must connect with higher level policies, capabilities and infrastructures. Grassroots innovators have to be sufficiently nimble to take advantage of windows of opportunity, like new funding programmes attached to shifting policy agendas, and cast themselves positively in the new light. But grassroots innovators find it extremely challenging to influence when and what form those opportunities take. A key challenge is to boost grassroots influence – local intelligence informing policy developments that further encourages diverse grassroots innovation (Roberts, 2005). Indeed, our central argument has been for a reconsideration of grassroots initiatives entwining the community action and sustainable innovation strands of higher-level sustainable development policy.

Reframing seeds of change

Technological innovation and community action are important strands of sustainable development that are rarely linked; the grassroots is a neglected site of innovation for sustainable consumption. Innovation literature describes the important role of niches in seeding transformations in wider socio-technological regimes, and in this chapter these ideas were adapted to grassroots activities for sustainable development in the social economy, and the implications of this conceptual development were discussed. The characteristics of grassroots innovations were described, demonstrating the links with New Economics theory and practice, in opposition to mainstream regimes; the benefits and challenges for these niches were discussed in terms of intrinsic and diffusion outcomes. Applying innovation theory to grassroots initiatives offers a new conceptual lens through which to view these activities, and chapters that follow put these theories to the test by examining a series of grassroots innovations for sustainable consumption. In each case, the niche characteristics are identified, and the implications of this alterity are considered, highlighting opportunities and obstacles for wider diffusion of niche ideas and practices for sustainable consumption.

5
Sustainable Food: Growing Carrots and Community

> Changing food purchase habits can dramatically alter the climate change impacts of our lifestyle. In its implications, it is comparable with the decision to abandon air travel (Goodall, 2007: 230).

It could be said that local organic food is flavour of the month. In recent years there has been a growing interest in the phenomenon of 'alternative agro-food networks', and locally-sourced organically-produced food has been proposed as a model of sustainable consumption. The claimed benefits include rural regeneration, livelihood security, cutting food miles and carbon dioxide emissions from transport, social embedding, community-building, and increasing connection to the land. Consequently, the recent revival of localised food supply chains and the rise in demand for specifically local organic produce has been described as a move towards a more sustainable food and farming system in the UK, and has driven the explosion of a grassroots movement of niche direct marketing outlets (farmers markets, farm shops, and veggie box subscription schemes) where consumers buy directly from growers. Are these consumers actively engaged in creating new food supply chains based upon alternative values to the mainstream? Are they constituting more sustainable systems of food provision, and if so, what is the potential for these niche initiatives to influence the wider regime of mainstream supermarket-dominated food supply chains? Are consumers expressing ecological citizenship, and is this the source of their motivation to consume more sustainably? With these questions in mind, this chapter examines the sustainable consumption rationales for local and organic food systems, and takes a close look at one such initiative in order to assess its effectiveness at delivering more sustainable food. It then investigates the scope and potential for this particular green niche to diffuse, with a study of the implications of mainstream supermarkets adopting niche local food practices.

The rationale for organic and local foods

Organic production refers to agriculture which does not use artificial chemical fertilisers and pesticides, and animals reared in more natural conditions, without the routine use of drugs, antibiotics and wormers common in intensive livestock farming. Consumer demand for organic produce has risen enormously over the last 15 years in the UK, growing from a niche activity to a mainstream consumption choice (Smith, 2006). Sales of organic products in the UK amounted to £1.213 billion in 2004, a rise of 11% on the previous year (Soil Association, 2005b), and the most commonly cited reasons for consuming organic food are: food safety, the environment, animal welfare, and taste (Soil Association, 2003). Simultaneously the area of land within the UK certified (or in conversion) for organic production has risen dramatically: in 1998 there were under 100,000 hectares and by 2005 this had risen to 690,000 hectares (DEFRA, 2005c). In its efforts towards a sustainable food and farming system, the UK government has pledged to support the growth of organic farming by promoting organic food in schools and hospitals, providing cash for farmers to help them transfer to the new organic farming system, recognising and valuing the social and economic benefits of organic farming, as well as environmental gains, and encouraging supermarkets to source more organic food from the UK (DEFRA, 2002). The main environmental rationale for organic agriculture is that it is a production method more in harmony with the environment and local ecosystems. Proponents claim that by working with nature rather than against it, and replenishing the soil with organic material, rather than denuding it and relying upon artificial fertilisers, then soil quality and hence food quality will be improved and biodiversity will be enhanced. A second rationale for organic food is to protect individual's health by avoiding ingestion of chemical pesticides (Reed, 2001) Additionally there are increased economic and employment benefits from organic farms compared to conventional farms (Maynard and Green, 2006). Supermarkets currently dominate the organic retail sector, and they have been quick to respond to the changing demand for local and organic produce. Between 2003 and 2005, the proportion of key organic staples sold in the eight main UK supermarkets which were UK-sourced has risen from 72% to 82% (Soil Association, 2005b).

One consequence of the growth in organic farming from a green niche to a mainstream system of production, highlighted by Smith and Marsden (2004), is the 'farm-gate price squeeze' common within conventional agriculture, which limits future growth and potential for rural development. Farmers keen to diversify into organic production as a means of securing more sustainable livelihoods in the face of declining incomes within the conventional sector are confronted with an increasingly efficient supermarket-driven supply chain which sources the majority of its organic produce from overseas. Currently 56% of organic produce eaten in the UK is imported, and 75% is sold through supermarkets (Soil Association, 2005a), representing an overwhelmingly mainstream distribution channel. A key challenge for small organic producers is therefore to create new systems of provision to bypass the supermarket supply chain, and organise in such a way as to wield sufficient power in the marketplace.

Organics has until the 1990s been a niche environmental interest, expressing a desire to bypass intensive agriculture and return to small-scale production, and grow a new sense of connection with the land, through a concern for the authenticity and provenance of the food we eat (Ricketts Hein *et al.*, 2006; Holloway and Kneafsey, 2000). In other words, it is a social as much as a technological innovation (Smith, 2006). As such, it has been representative of a movement towards the (re)localisation or shortening of food supply chains, and explicitly challenges the industrial farming and global food transport model embodied in conventional food consumption channelled through supermarkets (Reed, 2001). Localisation of food supply chains means simply that food should be consumed as close to the point of origin as possible. In practice, this will vary from produce to product, and the construction of 'local' is both socially and culturally specific, and fluid over time and space (Hinrichs, 2003); in the UK, consumers generally understand 'local' to mean within a radius of 30 miles or from the same county (IGD, 2003). A recent poll found that 52% of respondents with a preference want to purchase locally grown food, and another 46% would prefer it grown in the UK (NEF, 2003), and the growth of direct marketing, regional marketing and other initiatives has supported this turn towards 'quality' and 'authentic' local food (Holloway and Kneafsey, 2000; Murdoch *et al.*, 2000).

The principal environmental rationale for localising food supply chains is to reduce the impacts of 'food miles' – the distance food travels between being produced and being consumed – and so cutting the energy use and pollution associated with transporting food around the world. Much transportation of food around the globe is only economically rational due to environmental and social externalities being excluded from fuel pricing (Jones, 2001). This results in the sale of vegetables and fruit from across the globe, undercutting or replacing seasonal produce in the UK. Pretty (2001) calculates the cost of environmental subsidies to the food industry, and compares the 'real cost' of local organic food with globally imported conventionally produced food. He finds that environmental externalities add 3.0% to the cost of local-organic food, and 16.3% to the cost of conventional-global food. Furthermore, by avoiding the use of petrochemical-derived fertilisers (producing nitrous oxide, a potent greenhouse gas), the climate change implications of organic production are significantly below that of conventional agriculture. Goodall (2007) estimates that UK annual per capita CO_2e emissions related to food are 2.1 tonnes, from which the main contributing factors are artificial fertilisers, methane from animals and slurry, landfill gas from rotting food, food and drink manufacturing and processing, and packaging manufacture. This figure can be reduced to 0.35 tonnes through a radical shift in consumption patterns, switching to locally produced, organic food, and adopting a vegan diet.

A report commissioned by the UK government to investigate the utility of the 'food miles' concept for sustainable production and consumption finds that the direct environmental, social and economic costs of food transport are over £9 billion each year, of which over £5 billion are attributed to traffic congestion (and the value added by the agricultural sector is £6.4 billion and by the food and drink manufacturing sector £19.8 billion) (Smith, Watkiss *et al.*, 2005). Although some deeper examinations of the food miles idea exposes contradictions and counter-intuitive complexities in terms of life-cycle energy use, food production and transport (see for example

Born and Purcell (2006) and Schlich and Fleissner (2005)), the 'food miles' concept has nevertheless become an easily communicated idea to rally local food activists, and is here employed for its utility in capturing consumer motivations.

In recent years the major supermarkets have increased the availability – and visibility – of local produce within their stores. Observation in a local store shows that produce such as locally-brewed beer, honey and preserves, biscuits, cheese and sweets are grouped into a separate 'locally produced' section of the supermarket, and sold in folksy packaging in a clear attempt to win back customers who might otherwise buy from a farmers market or other direct marketing outlet (see Figure 5.1). Asda supermarket (the second largest in the UK with 17% market share and 258 stores, and owned by US giant Wal-Mart) is emblematic of the major supermarket chains. In a response to the growing demand for local produce, Asda introduced a local produce section in 2001 and now sells 2500 regionally-produced items from 300 local producers in its stores (for instance Norfolk beers and ales in its Norfolk stores), with an aim to achieve 2% annual turnover from

Figure 5.1   Locally-produced goods on sale in a UK branch of Wal-Mart Asda

local produce by 2008 (Mesure, 2005). The supermarket won the BBC Food and Farming award for best retailer in 2005, on the basis of its policy for supporting local speciality producers (AMS, 2005). Asda said it is 'actively encouraging local growers and farmers to deliver produce directly to their local store instead of supplying via a regional depot, ensuring it is fresher, has travelled far fewer food miles, and has a longer shelf life' (AMS, 2006). This move is tapping directly into shoppers' concerns about supporting local economies and farmers, as well as offering improvements in freshness and taste and a perceived local authenticity (Padbury, 2006) which many have criticised the supermarkets for eroding (Corporate Watch, n.d.).

In addition to this environmental rationale, there are also social and economic reasons to embrace re-localised food supply chains within a framework of sustainable consumption. In direct contrast to the globalised food system which divorces economic transactions from social and environmental contexts, the 'New Economics' favours 'socially embedded' economies of place. This means developing connections between consumers and growers, boosting ethical capital and social capital around food supply chains, educating consumers about the source of their food and the impacts of different production methods, creating feedback mechanisms which are absent when food comes from distant origins, and strengthening local economies and markets against disruptive external forces of globalisation (Norberg-Hodge *et al.*, 2000). Indeed, rather than being eroded by the demands of globalisation, these diverse embedded food networks are now flourishing as a rational alternative to the logic of the global food economy (Whatmore and Thorne, 1997).

Furthermore, there is a strong case that localised food networks make a significant contribution to rural development, help mitigate the crisis of conventional intensive agriculture, and have the potential to mobilise new forms of association which might resist the conventional price-squeeze mentioned above, through the development of new relationships and methods of adding value (Renting *et al.*, 2003). Direct marketing of local and/or organic produce through farm shops, farmers markets and box schemes has been proposed as a more sustainable, alternative infrastructure of food provision, for economic, social and environmental reasons (FARMA, 2006; Taylor *et al.*, 2005). Recent research by the New Economics Foundation found that street produce and farmers' markets made a major contribution to local economies, and provided access to fresh fruit and vegetables at prices significantly lower than nearby supermarkets (Taylor *et al.*, 2005). This is demonstrated in a study of food supply chains in Norfolk which found that the motivations for many growers to sell locally included 'taking more control of their market and [becoming] less dependent on large customers and open to the risk of sudden loss of business' (Saltmarsh, 2004, chapter 3). Many of these growers had previously supplied to supermarkets and for these farmers, direct marketing was a means of stabilising incomes and reducing vulnerability.

More evidence of this move towards alternative systems of food provision is reported by a study finding that 51% of organic growers in the UK were planning to work cooperatively with other farmers, to increase their market share and improve resilience against external economic shocks (ADAS, 2004). Indeed, sales of organic and local produce through these alternative direct marketing channels have grown rapidly. The number of farmers markets in the UK has grown rapidly since the first was established in Bath in 1997, to over 500 in 2006 (FARMA, 2006) and farmers' markets sales in 2004 amounted to £200 million, of which about 10–15% of stallholders sold organic produce, accounting for £25 million, a 21% increase on the previous year. Total sales of organic produce through direct marketing rose by 33% in 2004 to 12% of the entire food market share, a total of £146 million. There are an estimated 379 vegetable-based organic box schemes in the UK, and a further 97 meat-based schemes, and the sales from organic box schemes in 2004 was £38.5 million. Reflecting this growth in market share, the supermarket share of the organic retail market has fallen from 80% in 2003 to 75% in 2004 (Soil Association, 2005b). In addition to insulating farmers, localisation also builds up the local economy by increasing the circulation of money locally (the economic multiplier). In a study of the economic impact of localised food supply chains, Ward and Lewis (2002) found that £10 spent with a local grower circulated two and a half times locally and was worth £25 in the local economy. This compares to £10 spent in a supermarket which leaves the area quite quickly, resulting in a multiplier of just 1.4, meaning it was worth £14 to the local economy.

Localism is not uncritically embraced, however, within the New Economics. Localisation can be a reactionary and defensive stance against a perceived external threat from globalisation and different 'others' (Hinrichs, 2003; Winter, 2003), and the local can be a site of inequality and hegemonic domination, not at all conducive to the environmental and social sustainability often automatically attributed to processes of localisation by activists. Indeed, Thompson and Arsel (2004) describe such uncritically pro-localisation consumers as 'oppositional localists', marked by their attribution of only positive characteristics to small-scale local organisations and businesses, and their wholesale rejection of globalised business. Their research points to the need to be objective about the motivations of consumers, and the underlying values they represent. But localism also raises questions of 'sustainability for who?', as the nascent desire for locally produced food in developed countries inevitably impacts upon the economic and social destinies of food-exporting developing countries. In these cases it may be that ecological citizenship which calls for cutting material consumption and hence a reduction in globally transported foodstuffs, is in conflict with a particular type of global citizenship which holds that participation in international trade is the most effective route to sustainable development for poorer countries. New Economists argue for a globalised network of local activism which addresses the economic and social needs of developing countries reliant upon food exports, and which prioritises fair trade for products which cannot be produced locally. Banana Link is one such organisation, which seeks to build solidarity links between UK consumers and retail workers and Central American banana growers and farm workers struggling to improve working conditions and local environments, while simultaneously lobbying at the international level to improve the terms of trade (Banana Link, 2003). Hence a reflexive localism offers ecological citizens the opportunity to forge both local and global alliances with progressive actors at the local level and consciously avoid the negative associations of defensive localism (DuPuis and Goodman, 2005).

Evaluating a grassroots sustainable food initiative

The sustainable consumption rationales for local and organic food networks are manifold and wide-ranging, as the discussion so far has shown. But how effective are such practices at achieving their goals, and what is the scope for grassroots niche practices in sustainable food to influence mainstream provisioning? To answer these questions, the findings of an investigation into an organic food producer cooperative in the UK are now presented.[1] This East Anglian organisation, named Eostre Organics (pronounced 'easter', and named after the Anglo-Saxon goddess of regeneration), aims to build a 'fair, ecological and cooperative' food system, and sells to local businesses and hospitals as well as through market stalls and weekly subscription boxes of mixed vegetables and fruit delivered direct to consumers throughout the region. Eostre won the Local Food Initiative of the Year award in the Soil Association's Organic Food Awards in 2003, given to the business or venture considered to have shown most 'innovation and commitment in making good food locally available' (Eostre Organics, 2004). It is therefore emblematic of the model of sustainable food elaborated above, and furthermore is considered an exemplar of its type.

Eostre Organics

Eostre's origins lie within Farmer's Link, a Norfolk-based NGO which was inspired by the Rio Earth Summit in 1992 to improve the sustainability of farming in developed countries, and making solidarity links with UK farmers. In 1997, it set up East Anglia Food Link (EAFL) to promote conversion to organic production in the region. EAFL's vision is one of localism – building direct links between farmers and consumers to create more sustainable food supply chains and benefit local economies and communities (EAFL, 2004). EAFL developed links with European organic growers and was inspired by the strength and

---

[1]The research was a multi-method study carried out during the spring of 2004, and consisted of site visits to Eostre's headquarters and market stall, interviews with organisers and staff, documentary analysis of their web site and newsletters to ascertain the scope and nature of activities, objectives and values. This was complemented by two self-completed customer surveys: the first survey of market stall customers achieved 65 responses out of 110 distributed over a two-week period (59%); the second surveyed the 252 customers of three weekly box schemes supplied by Eostre (79 responded, giving a response rate of 31%). The surveys asked about motivations for, and experiences with consuming local organic food, and are considered together here (overall response rate 39%) unless specified otherwise. There were both closed- and open-ended questions in order to elicit the respondent's own interpretations and meanings of their actions and the discourses they used to explain them. Qualitative analysis was used to code and analyse these responses, alongside quantitative analysis of other data.

growth of producer cooperatives, and persuaded local organic growers who were already intertrading informally, to adopt a formal cooperative structure to develop new markets and help grow the member businesses. Eostre was established in 2003 with a DEFRA Rural Enterprise Scheme grant, with nine members, seven associate or prospective members including one overseas member: the El Tamiso organic producer cooperative in Padua, Italy, which itself comprises over 50 businesses. Eostre Organics is a food business with a mission: its charter states:

> Eostre is an organic producer co-operative supplying fresh and processed organic food direct from our members in the East of England and partner producers and co-operatives from the UK and Europe. Eostre believes that a fair, ecological and co-operative food system is vital for the future of farming, the environment and a healthy society. Direct, open relationships between producers and consumers build bridges between communities in towns, rural areas and other countries, creating a global network of communities, not a globalised food system of isolated individuals (Eostre Organics, 2004).

Its specific aims include: to supply consumers of all incomes high-quality seasonal produce; to encourage cooperative working among its members and between the co-op and consumers; transparency about food supply chains; to source all produce from UK and European regions from socially responsible producers and co-ops promoting direct local marketing, and from fair trade producers outside Europe; to favour local seasonal produce and supplement (not replace) with imports; to minimise packaging, waste and food transport; to offer educational farm visits to raise awareness of the environmental and social aspects of local organic production (Eostre Organics, 2004). From these objectives, it is clear that Eostre is strongly supportive of the New Economics model of sustainable consumption, which favours re-localisation, reducing environmental impacts and ecological footprints, and that there are clear expressions of ecological citizenship values here too. How do these translate into practice?

In the Eastern region of the UK, farm employment has fallen from 66,305 in 1990 to 49,409 in 2003, a drop of 25% (DEFRA, 2003a), and Eostre aims to tackle this decline in rural employment by supporting small growers. Between the nine local members, Eostre accounts for 1055.8 ha of diverse farmland, including 1.6 ha (with a quarter of this under glass) to 48.6 ha of farmland on rich fenland peat, to 445.2 ha of arable farmland and grazing pasture. The average farm size of Eostre members is 117.3 ha, though most are much smaller than this: three are less than 5 ha, and the median is 24.3 ha. In comparison with the agricultural sector in the region where the average holding is 73.9 ha, most of Eostre's farms are very small (DEFRA, *ibid.*) and they are mostly 100% organic. Normally, this is a problem for growers seeking to supply local markets, as stability of supply cannot be guaranteed. However, through collective organisation, Eostre's members can achieve the scale required to penetrate such markets, for example by supplying market stalls and box schemes. Commercially, Eostre has been a success. The businesses of members grew over the first year or so that Eostre was operational, with an increase in sales of 70% over 12 months. The cooperative now supplies produce to 13 box schemes, 15 market stalls (including the UK's only full-time organic market stall on the general provisions market in Norwich city centre which has recently doubled in size, and weekly stalls in several market towns around Norfolk, plus monthly farmers markets), nine cafes, pubs or restaurants and 12 shops. Inroads have been made into public sector catering, through local schools, hospitals and prisons.

The motivations of Eostre's consumers were surveyed to explore whether and to what extent ecological citizenship values played a part in their decision to purchase food from Eostre. Survey respondents were asked why they chose to purchase from Eostre. The responses fell into four main groups: environmental, economic, social and personal benefits (see Table 5.1). The two most commonly-given reasons were both environmental, namely because respondents thought local organic food was better for the environment (94% of respondents gave this answer) and to cut packaging waste (85%). The next most frequently-given motivations were to cut food miles (another environmental driver, with 84%) and to support local farmers (an economic factor, also 84%). Next, personal motivations of nutrition and safety were given by 80% and 77% of respondents. The most important social factor given was to know more about the origins of food and how it was produced (77%).

Clearly, the range of significant social, economic and environmental objectives expressed maps closely onto the sustainable consumption

Table 5.1   Consumers' motivations for purchasing from Eostre

	Ranking	% of customers (n=144)
Environmental benefits		
Better for the environment	1	94
To cut packaging waste	2	85
To cut food miles	3=	84
More diversity of produce varieties	11	33
Economic benefits		
Supporting local farmers	3=	84
Supporting a cooperative	8	70
Keeping money in the local economy	9	65
Social benefits		
To know where food has come from and how it was produced	7	76
Preserves local traditions and heritage	10	36
Enjoy face-to-face contact with growers	12	25
Demonstrates good taste and refinement	13	8
Personal benefits		
Organic food is more nutritious/ tastes better	5	80
Organic food is safer	6	77

Note: 3 = N third-equal
Source: Author's survey of Eostre customers.

goals of the organisation itself. This suggests that customers share the ecological citizenship principles of Eostre in seeking to develop sustainable food supplies through localised channels. And the consumers did seem to back up these principles with action: the average household expenditure on all food and drink of respondents was £71 a week; of this, over half (£37 or 52%) was spent on local or organic or fairly traded products (from all sources, not just Eostre). This represents a very significant use of consumption decision-making as political activity, and is far greater than the marginal expenditure found in other surveys. The Co-operative Bank (2007) found that while household spending on ethical products has doubled between 2002 and 2006, it is still only approximately £13 a week, suggesting that Eostre's customers are not representative of the general population. Rather, they may be described as a highly motivated group of ecological citizens, certainly conversant in discourses of sustainable consumption. They may have been introduced to these issues beforehand, or they may have learned about them as a result of interaction with Eostre, who adopt an educative, outreach role to inform and motivate consumers, through farm visits, newsletters, etc. In this manner, Eostre can be said to be actively nurturing ecological citizenship and simultaneously providing a means – and social context – for its expression.

In the next section Eostre is critically appraised in terms of its ability to deliver sustainable food, using the five sustainable consumption criteria developed in Chapter 3: localisation, reducing ecological footprints, community-building, collective action, and building new infrastructures of provision. The findings are summarised in Table 5.2.

Evaluating Eostre as a tool for sustainable consumption

Localisation

The principal aim of Eostre was to support the livelihoods of local organic producers within the region, by enabling them to serve local markets, and this aim has been achieved so far: Eostre saw a 70% increase in sales during the first year of operation, and has expanded its range of retail outlets. Indeed, an index of food relocalisation developed by Ricketts Hein *et al.* (2006) finds that Norfolk ranks 9th among the 61 counties of England and Wales. Consumers also value local producers highly, and 84% of the survey respondents said they chose Eostre because of a commitment to supporting local farmers. One consumer said: *'I value the fact that some of it is grown in Norfolk by small businesses whose owner and workers obviously care about the land, their customers and their social surroundings'*, and another stated *'I would like to see a return to seasonal fruit and veg, which we can only hope for is we support the smaller / local farms'*. Keeping money circulating in the local economy – by patronising locally-owned businesses – was a motivation for 65% of consumers who responded to the survey, for example *'we like to support local growers and local industry'*. The theme of self-reliance was also prominent, and one mentioned *'I like the idea of England being more self-sufficient and using our own good land to feed us all simply'*, and 36% of respondents wanted to preserve local traditions and heritage through supporting Eostre.

The localism and associated sense of connection between growers and consumers that this affords was important for many. This connection was facilitated through the personal contact provided by retail staff and the information they provided about the sources of food. For example, one customer explained '[the] source of food is more likely to be trustworthy and produced to a high standard. I like the traceability and accountability, as opposed to most supermarkets which are primarily accountable to their shareholders', and another wrote 'I like to know what I am eating and can trust the supplier that the food is fresh, local and natural'. Furthermore, Eostre organises educational farm visits so that customers can see where their food is grown, and publishes a regular newsletter which highlights sustainable food issues as well as offering recipe ideas and profiles of growers. In other words, there is a sense of community growing around this food network which nourishes its members, and enables them to participate as active members, and Eostre is attempting to promote and nurture the ecological citizenship which can then thrive in this meaningful social context.

Eostre's marketing officer explains that localising food supply chains is absolutely central to Eostre's operations: 'People are becoming very eco-aware, and one of the biggest issues in any ecological awareness has got to be food miles'. Indeed, food miles was a concept high in the minds of Eostre's customers when thinking about the localisation impacts. Eostre's marketing manager explains

Figure 5.2   'Think Global, Shop Local' sign on Eostre's market stall

'People are becoming very eco-aware, and one of the biggest issues in any ecological awareness has got to be food miles', and this is supported by the survey which found that 84% of survey respondents specifically aimed to reduce food miles through buying food from Eostre. Typical explanations included: 'If good, tasty food is available locally, it seems pointless to buy potentially inferior goods from a supermarket which have often been imported from across the globe', 'It cuts out the environmentally-destructive chain of transport from one end of the world to another' and 'It supports the local economy, reduces food miles, and enhances the local countryside'. However, at present consumers sometimes face a trade-off between local and organic attributes of their food, and must choose according to where their priorities lie, between conventionally-produced local food, and imported organic produce. In reference to organic food sold in supermarkets, Eostre's marketing officer claims that 'whatever benefits people gain from it being organic, they lose from the food miles it takes to get it here', which is a sentiment shared by many customers; one customer stated 'I don't believe [imported] organic is worth the food miles'. Yet the same argument can be made about some of Eostre's produce, as much is imported (from the Italian producer cooperative partner, and from other organic and fair trade suppliers around the world) in order to guarantee a wide range of produce all year round. For example, in May 2004 Eostre's market stall was selling organic broccoli from France, onions from Argentina and carrots from Italy, while conventionally grown local produce was available on neighbouring market stalls at considerably lower prices.

Some customers felt that they would prefer to see less imported produce, especially that which could be grown locally, and one stated 'sometimes there seems to be a lack of local produce, and I still think Eostre runs up quite a few food miles… what about stocking e.g. Norfolk asparagus or strawberries?'. This could be addressed by expanding the membership of local organic suppliers to provide a wider range of produce and so reducing reliance upon imported food, but the wider issue of how consumers should make choices between different 'sustainable' food choices is unclear in the absence of a food sustainability indicator which addresses the full range of issues involved – including the impacts of UK-based localisation on developing country producers.

Reducing ecological footprints

Much of the impacts for reducing ecological footprints has been covered alongside localisation, in the previous section, but there are further aspects to consider as well. A commitment to sustainable farming and food is evident in Eostre's mission statement above, and this is forcefully supported by their customers. Of the customers who responded to the survey, 94% stated that they bought from Eostre because they believed local and organic food was better for the environment. For example, one respondent replied '[buying local organic food] is important because we believe in sustainability regarding our environment, and we are committed to reducing our "'eco-footprint" in any areas we can', and another stated 'I feel I owe it to the Earth'. Other comments included: 'I am very concerned about the effects of pesticides and pollution on us and the environment', 'organic farming is better for wildlife' and 'I want to support a farming system that works within environmental/resource limits'. As these and previous statements suggest, the environmental factors being considered are farm-related (pesticide and fertiliser use), transport-related (food miles), and packaging-related (85% of respondents chose Eostre in order to reduce unnecessary food packaging). Another customer explained 'to me, it represents a more harmonious ecological balance between that which we produce, consume and waste'.

Community-building

In addition to strengthening the local economy and reducing environmental impacts, Eostre is also a community-building initiative. Local economic and community links are built up between farmers and consumers, and consumers gain a sense of connection to the land, through the personal relationships which develop. As one respondent explained, the appeal of Eostre was 'the sense of communal participation, starting from the feeling that we all know – or potentially know – each other, and continuing on through wider issues, both social and environmental', and another stated 'I feel that "connectedness" is important' while another reported that they liked Eostre because 'it's a cooperative; they are like-minded people'. These personal connections are developed in several ways: from face-to-face contact on the market stalls or with box-deliverers, and secondly through newsletters which share stories, recipes and news about the farms, and invite customers on educational farm visits. Three quarters

(76%) of those customers who completed the survey reported that they were motivated to purchase from Eostre because they liked to know where their food has come from, and a quarter (25%) specifically liked the face-to-face contact with growers. This sense of community is echoed by another respondent who favours local organic food because 'purchasing it links me with a part of the community which operates in a far healthier and more ethical way than the wider economic community', and another felt that 'organic food helps bring back small community living instead of alienated individuals feeling unconnected'.

Local organic food networks are builders of community and shared vision, and the Eostre market stall in Norwich is a good example of how this works: it is a convenient city-centre meeting point and source of information, open to everyone. The stall is decorated with leaflets and posters advertising a range of sustainable food and other environmental initiatives, for example anti-GM meetings, Green Party posters, alternative healthcare practices, wildlife conservation campaigns etc (see Figure 5.3). This correctly reflects the interests of

Figure 5.3   Eostre's market stall, an alternative green network hub

customers: 60% of respondents identified the Greens as the political party which best represented their views. But how socially inclusive is this community? Organic food is often dismissed as the preserve of an elite, on grounds of price, and claimed to be inaccessible to lower-income groups (Guthman, 2003). In fact many of Eostre's customers are from lower income brackets, broadly representative of the local populace. Comparing Eostre customers who responded to the survey, 14% of customers had a gross weekly household income of less than £150 (£7,800 a year), compared to 15% of the local population, and higher-income households were under-represented: only 17% of Eostre customers had household incomes of over £750 a week (£39,000 a year), compared to 23% of the local population (ONS, 2003). Only 8% of customers felt that eating organic reflected 'taste and refinement', suggesting that in this case, organic is not 'posh nosh'. With such a high proportion of low-income customers, Eostre is achieving its aim of making fresh organic produce available to all social groups.

Collective action

There are two ways in which Eostre is an expression of collective action for sustainable consumption. The first is through its structure – as a cooperative. Many of the farmers in the cooperative had previously sold organic produce to supermarkets, and had suffered from a drop in sales and prices during the recession in the early 1990s, as well as having a negative experience of dependency upon a single, distant buyer. This led some growers to seek greater control over their businesses by moving into direct marketing, and an informal inter-trading arrangement developed between a handful of small local organic growers, which formed the core of the cooperative. Eostre therefore aims to provide sustainable and stable livelihoods to its member growers, as a grassroots response to economic recession and vulnerability caused by a global food market – a local adaptation to globalisation in the food sector. By organising collectively, Eostre's members achieve the scale necessary to access markets which small growers cannot manage alone, for example being able to supply market stalls all year round. For example, one smallholding of under one hectare has been supported in developing new markets through collective box schemes and market stalls, and another farmer, who was struggling as a conventional fenland farmer, now has greater livelihood security as an organic producer within Eostre (Saltmarsh, 2004). This evidence indicates that there appears to be scope in this organisational structure and growth of direct marketing to avoid the limits to growth experienced by other parts of the developing organic sector identified by Smith and Marsden (2004) and Renting *et al.* (2003). The cooperative values were supported by customers: 70% of respondents said they chose to buy from Eostre in order to support a cooperative, and one stated 'I like that local organic farmers work together rather than competing against each other for profit'. Another customer commented 'I like the knock-on effect of supporting local cooperative and organic farmers', and another stated 'I object to [supermarkets'] attitude to suppliers (i.e. squeeze them to keep the prices low)'. This empowerment through daily private decision-making with political implications is a core aspect of ecological citizenship, and the evidence suggests that Eostre enfranchises its customers with a feeling of political agency which fulfils their need for expression and activism.

The second collective action impact is through Eostre's inroads into public sector catering through small-scale initiatives such as providing food for a primary school kitchen, and supplying the local hospital visitor's canteen. These were important first steps, albeit an uphill struggle against the ingrained habits and beliefs among public sector catering managers, and institutional barriers such as the lack of a kitchen to feed patients in hospitals (cook-chill food being the norm). However, the changing public agenda on school meals as a result of Jamie Oliver's 'School Dinners' TV programme has thrust local organic food provision into the limelight, and Eostre and parent NGO East Anglia Food Links have been identified as pioneers with important lessons to share. Currently heads of catering from seven of the ten East of England Local Education Authorities have agreed to work together with EAFL, on a programme of work to increase the use of sustainable and local food in their school meals (EAFL, 2005).

The government currently advises schools to consider alternative suppliers, but organic local food will not get into schools on any large scale until there are government directives instructing that schools must use organic local produce. Eostre felt that the existing supply chains have been in place for so long, there was no incentive to change them, and there was resistance in public sector organisations to new approaches to food. In particular, they felt that organics were still seen as 'alternative' to many people in positions of power, and that a pro-active push from government would be needed in order to achieve significant changes in these institutions. Introducing localised food supply chains into this institution would require changes in the infrastructure within these institutions. Provision (or not) of a kitchen to feed patients is a decision made at the planning stages of a building project, and has implications for patients health and wellbeing, as well as for the options available for managers to implement alternative arrangements (SDC, 2004). If these obstacles could be overcome and public sector infrastructure put in place to enable this type of change, there is a huge potential for initiatives such as this to transform local markets, particularly with regards to local organic food supply chains, as well as provide strong leadership from government about desirable consumption patterns.

Building new infrastructures of provision

The successes which Eostre has achieved in the previous four categories add up to more than the sum of their parts: together they comprise the seeds of a new system of food provision, based upon cooperative and sustainability values (such as fair trade), and bypassing supermarkets in order to create new infrastructures of provision through direct marketing. Furthermore, their consumers actively support this activity, and many commented on how they enjoyed the opportunity to avoid supermarket systems of provision, for example: 'I think that supermarkets are distancing people from the origins of food and harming local economies; I try to use supermarkets as little as possible', '[Eostre is] an alternative to a system which rips off producers, the planet etc', 'I believe in a local food economy' and 'I don't want supermarket world domination, extra food miles, packaging, and middle people making money!'.

The consumer values expressed in these new institutions are quite different to those in mainstream systems of provision. For example, customers appear to be internalising calculations about social and environmental costs of conventional food production and transport, in order to respond to more sophisticated and inclusive price incentives than those in the marketplace. One stated 'I like to pay the "real cost" for my food' and another commented 'While not always as cheap as supermarket produce, I am more comfortable knowing that a greater proportion of my money goes to the primary producers'. A

Table 5.2   Evaluating a local organic food cooperative as a tool for sustainable consumption: key findings

Sustainable Consumption Indicator		Eostre Organics
		Award-winning organic producer cooperative based in Norfolk, East Anglia. Supplies market stalls, box schemes, shops and restaurants.
Localisation	👍	Improving the security of livelihoods for growers; keeping money circulating in the local economy; better feedback between producer and consumer; increasing a sense of connection with the land; promoting local food, so cutting food miles.
Reducing Ecological Footprint	👍	Cutting food miles and associated energy use; organic production avoids use of artificial pesticides, fertilisers and so is better for the environment and lower-carbon; reduced packaging;
Community-building	👍	Forging links between growers and consumers; developing a sense of community; growing social capital around food networks; accessing low-income consumers.
Collective Action	👍	Cooperative structure shifts the incentive and reward systems for producers; influencing public provision through schools and hospitals.
New Infrastructures of Provision	👍	Developing a value-based cooperative structure; avoiding mainstream food supply chains (supermarkets); prioritising local, organic, seasonal produce which runs counter to the norm of year-round supply; enjoying limited choice of produce.

second difference is the embracing of seasonality and acceptance that certain foodstuffs will not be available for several months of each year. In addition, subscribers to the box schemes do not even have a free choice over what food they will receive, instead being given a box of mixed seasonal fruit and vegetables each week – one likened the inherent surprises to 'having a Christmas present every week! I never know what the box will contain, it's a challenge to my cooking skills!', and others echoed the pleasure in adapting to seasonal availability. While a temporal lack of produce variety might be seen as a major failing in mainstream systems of food provision (the vision of empty supermarket shelves inducing panic!), within this infrastructure it is welcomed as an indicator of connection with the seasons and locality. One customer remarked 'I reject the ethos of the supermarket that all products should be available all year round. I enjoy the seasonal appearance of purple sprouting broccoli, asparagus, etc', and many comments referred to creating new sustainable food systems, confirming the notion that Eostre is beginning to create new provisioning institutions and socially 'embedding' economic relationships around localised food supply chains and networks (Whatmore and Thorne, 1997).

Diffusing the benefits of sustainable food niches

We have already seen that Eostre is an innovative green niche, a grassroots response to the mainstreaming of organic production which expresses values counter to the regime. The sustainability transitions management literature described in the previous chapter outlined three possible routes for strategic niches to diffuse and influence regimes. These are replication, upscaling and translation of ideas to mainstream contexts. The first of these is already being seen across the UK, with the rapid growth in the number of farmers' markets, box schemes, and other alternative food networks such as cooperative networks and solidarity-based organisations such as Riverford Organics, a producer cooperative like Eostre which supplies home delivery box schemes. The second route, upscaling, is also evident in the development of some very large-scale box schemes such as Abel and Cole (www.abelandcole.co.uk) who have 60 farm suppliers, deliver widely across the south of England, and achieve far greater economies of scale than Eostre could manage. Similarly, Riverford Organics is part of a wider franchise network covering the whole of the south and east of England, offering a consistent and customer-friendly online interface for consumers, while tapping into the resources of many diverse suppliers. Each of these routes to diffusion of the niche innovation is growing successfully at present, but faces a powerful threat: super-market competition. Now that supermarkets are embracing local as well as organic produce, how will this affect Eostre's development? The encroachment of mainstream retailers into the market for local and organic produce is an indication that niche practices are being adopted by the regime, but as the following discussion shows, it fundamentally threatens the existence of the niche itself.

While Eostre's customers reported that half their weekly food and drink expenditure was on local, organic or fairly traded products, three quarters (75%) of the respondents reported that they bought some of this produce from supermarkets, despite a general antipathy towards the mainstream supermarket system *per se*. One respondent remarked 'I generally find supermarkets unappealing and feel [it is valuable to do] anything I can to prevent the homogenisation of food'. This paradoxical behaviour indicates that for most people, food provisioning systems are not an all-or-nothing choice, but rather a plurality of approaches and systems reflecting perhaps the trade-offs between affordability, accessibility and ethics. How then does Eostre compete with mainstream supermarket provisioning?

Eostre's marketing officer is certainly wary about the impact of becoming more mainstream as a business, and indicates a prefer-ence to remain within a niche in order to protect core values and practices:

> I'm not sure that we are really aiming our produce at mainstream markets, as an ethical/environmental company it's as important if not more important that we adhere to our beliefs in sustain-ability, both environmentally and financially. If these issues let us into mainstream outlets then that's great if not we will prob-ably continue to seek out the more peripheral customers.

In order to uncover the underlying threats and opportunities for direct marketing as opposed to supermarket provisioning, the survey asked Eostre's customers open-ended questions about their views on direct marketing versus supermarket channels of food provisioning; these were grouped into categories and are shown in Tables 5.3 and 5.4, and some of the statements made are presented here to illustrate the points. Here the views of box scheme customers and market stall cus-tomers are disaggregated in order to better understand the consumers' views about particular aspects of the type of direct marketing they engage with.

When asked to list the advantages of purchasing from Eostre compared to through supermarkets, customers responded to the survey with a set of issues which bore a striking similarity between stall and box scheme customers (summarised in Table 5.3). For stall customers, the main ones are: supporting local businesses (51% of respondents); ethical consumerism and avoiding supermarkets on principle (38%); reduced packaging waste (35%) and cutting food miles (22%). For box scheme customers, the principal factors are: again, supporting local businesses (54%); better quality produce (42%); convenience (31%); and cutting packaging (30%). So consumers are making a strong statement that purchasing from a supermarket was not equivalent to buying from Eostre, as it meant losing some of the qualities they cherished – and the most important of these was localism. Of the top six advantages mentioned by each user group, the only issues specific to each direct marketing route were the convenience of getting a weekly delivery for box scheme customers, and the friendly atmosphere on the market stall.

Questions about the disadvantages of purchasing through Eostre when compared to through a supermarket (shown in Table 5.4), provoked fewer responses, and a much narrower range of factors was suggested by survey respondents. Interestingly, the most commonly-cited disadvantage for each group of customers was directly related to the

Table 5.3   Consumers' perceptions of the advantages of direct marketing compared to supermarket provisioning

Box Scheme Customers (n=74)	%	Market Stall Customers (n=63)	%
Supporting a local business	54	Supporting a local business	51
Better quality produce	42	Ethical shopping / not a supermarket	38
Convenience	31	Reduced packaging	35
Reduced packaging	30	Reduces food miles	22
Ethical shopping/not a supermarket	24	Friendly atmosphere	21
Reduces food miles	23	Better quality produce	21

Source: Author's survey of Eostre customers, multiple responses allowed, hence totals exceed 100%.

Table 5.4   Consumers' perceptions of the disadvantages of direct marketing compared to supermarket provisioning

Box Scheme Customers (n=50)	%	Market Stall Customers (n=50)	%
Limited or no choice	50	Less convenient/accessible	56
Higher price	20	Higher price	26
Limited range	18	Lower quality produce	20
Lower quality produce	6	Limited range	10

Source: Author's survey of Eostre customers, multiple responses allowed, hence totals exceed 100%.

provisioning route chosen. Eostre's market stall customers felt that the principal drawbacks of sourcing organic food through Eostre compared to supermarkets were related to convenience and accessibility (56% of stall customer respondents cited this problem). This included limited opening hours (the stall is open from 9am till 5pm, 6 days a week), and the difficulty of carrying heavy shopping bags back from the city centre. Higher prices was the second-most often reported disadvantage of Eostre over supermarkets (26%), followed by poorer quality of produce (20%). In contrast, box scheme customers felt that the limited choice and inability to select produce was the biggest drawback compared to using a supermarket (50% gave this response) although many said that they personally did not find it a problem. Price was again the second-most cited disadvantage (20%), followed by an acknowledgement that the range of produce available was more limited than a supermarket would offer (10%).

These preferences reveal the strengths and weaknesses of direct marketing niches compared to mainstream supermarkets for a specific group of committed direct marketing consumers. While the major reason to choose direct marketing over supermarkets is related to supporting local businesses and strengthening the local economy, and could not easily be challenged by international supermarket chains, there are other niche practices which might be incorporated into the mainstream food supply chain. These might include measures to reduce the packaging in supermarket food, or to source more produce locally, and may win the custom of less ideologically-committed consumers. Conversely, by addressing the stated disadvantages that Eostre's customers report, it is conceivable that supermarkets might capture some of Eostre's market share (or indeed, prevent it from expanding to a broader customer base) if they can provide fresh organic or local produce that is cheaper, more diverse, better in quality, and/or more conveniently available.

If this happens, and current developments in supermarket provisioning suggest that it is a goal of the mainstream suppliers to do so, it would imply that alternative food networks may be no more than a transitory phase in the adaptation of mainstream systems of provision to the demands of green and ethical consumers, but that this adaptation process results in a dilution of the radical transformative aims of those innovative system-builders. This process has been observed within the organics movement, as mainstream incorporation of organics has concentrated on the technical specifications of production systems, while neglecting the deeper social change inherent in the organics movement's original aims:

> Organic equivalents of highly processed conventional food products appeared on supermarket shelves, e.g. frozen ready-meals, fizzy drinks. Organic produce was not transforming the food regime; it was simply a new, high value ingredient threading its way into conventional food socio-technical practices (Smith, 2007: 422).

However, this translation of organics from niche to mainstream food production system has been accompanied by a splintering of the organic movement, ensuring that a renewed radical niche exists to continue to push for system-wide change, concentrating on social contexts and community action (*ibid*). This research with Eostre finds that some of the motivations given for purchasing from a direct marketing initiative could, conceivably, be expressed through purchasing from supermarkets: certainly if organic certification of produce is the principal concern, then supermarket provisioning meets that need more than adequately. But other issues are not so easily transferred into the mainstream supply chain: supporting a cooperative, keeping money in the local economy, having face-to-face contact with growers and increasing one's connection with the source of one's food are all aspects which appear to be the antithesis of the supermarket model.

Ecological citizenship and sustainable food innovations

This chapter began with a question: could ecological citizenship be a new force to motivate sustainable consumption, and are such motivations expressed through purchasing food from local organic food networks? Having reviewed and evaluated the activities and discussed the motivations of the participants of one such network – Eostre Organics – three things become apparent. The first is that Eostre – as envisaged and practised by its creators and users – is a niche, grassroots-based sustainable food initiative rather than a mainstream project. Furthermore, it is effectively developing new social and economic institutions for sustainable consumption, and successfully addresses all five of the sustainable consumption criteria. The alternative model of sustainable consumption demands localisation and re-embedding the economy within social networks, and Eostre is a good example of how this might work in practice. It uses food as a mechanism for community-building and social cohesion, while delivering sustainable rural livelihoods and a channel for the expression of alternative values about society, environment and the economy. Second, the values and principles expressed by both creators and users of this local organic food network are strongly resonant with ecological citizenship, and a powerful environmental ethic is a significant – if not primary – motivation for many of the participants. They sought to express preferences which were at odds with market price signals, they demonstrated a clear moral commitment to justice and fairness in trading relationships, to reducing ecological footprints through localising food systems and reducing packaging waste, and sought to make links of solidarity between producer and consumer, regardless of geographical distance. Furthermore, many participants saw their everyday consumption decisions as being deeply political, and enjoyed the expression of values – and small changes brought about – as a result of this quotidian political activity. There was a strong sense of participation in an alternative infrastructure of provision based on different values to the mainstream, and consumers felt actively engaged in creating and supporting this system. And third, the relationship between ecological citizenship, local organic food networks and sustainable consumption is more complex and sophisticated than might first appear. It does appear that ecological citizenly values motivate individuals to purchase their food from local organic food networks, in order to achieve sustainable consumption objectives. But there is an additional causal relationship to consider: namely the influence that local organic food networks have on promoting ecological citizenship and developing informed, educated communities around food

– through education, outreach, literature, farm visits, web sites, etc. – and so both nurturing the ethics of ecological citizenship and then providing a means for their expression. Indeed, many participants used the language and vocabulary of ecological citizenship when explaining their motivations: reducing ecological footprints and cutting consumption were commonly cited, in addition to more personal health and safety reasons.

In terms of Dobson's theory of ecological citizenship, we can say that Eostre and its consumers are behaving as 'good ecological citizens', and this citizenship model has proved a valuable analytical tool to understanding their values and motivations in a way which conventional theories of citizenship – neglecting citizenly activity taken in the private realm, and that which begets responsibilities to people beyond the nation state – do not. Similarly, it bridges the analytical divide commonly placed between 'citizen' and 'consumer' preferences (Sagoff, 1988), to describe activities which derive from citizenly urges, but which take place in the consumer sphere – the bedrock of sustainable consumption. Having tested the theory against an empirical study, ecological citizenship is therefore found to be a valuable theoretical model, and may indeed be a useful route to achieving a transition to deeper, 'alternative' sustainable consumption through a personal commitment to global environmental and social justice rather than top-down regulation.

Taking a wide perspective on sustainable food, and the potential for grassroots niche innovations such as Eostre, the implications of these findings for sustainable consumption are profound: while supermarkets offering organic and local produce may capture some of the consumer market for these goods, they remove support for other sustainability-related aspects of their production which are held as equally valuable by direct marketing consumers. Such developments attract customers with convenience, choice and low price (Padbury, 2006), but do not respond to the need for community-building, personal interactions between farmer and consumer, and for strengthening local economies and livelihoods against the negative impacts of globalisation, which consumers also express. Consequently, the beneficial impacts of local and organic food consumption are reduced in scope, and the potential for alternative food networks such as local direct marketing initiatives to expand and increase their influence on food provisioning systems is reduced. Therefore there is an urgent need for policymakers and analysts to recognise and demonstrate the wide-ranging benefits of direct marketing initiatives for sustainable consumption, to raise awareness of the interconnected social, economic and environmental issues surrounding food provisioning systems, and to support initiatives seeking to construct alternative infrastructures of provision.

While the market for organic food has grown in recent years, there are still large stumbling blocks to overcome. These are, first, the fact that local organic produce costs more than imported conventionally grown food, and second, difficulties expanding into supplying the public sector despite government recommendations to hospitals and schools to source food supplies locally and organically where possible. In these cases, infrastructure weighs against local sustainable food supplies (for example, the largest hospital in Norfolk does not have a kitchen to feed its patients) as well as social acceptability – organic food is still regarded as cranky by those in authority – and so the potential growth of sustainable food consumption through public procurement is hampered (Morgan and Morley, 2002). While market signals continue to misdirect, small groups of committed ecological citizens form a niche following their values rather than their purses. If pricing were corrected, for example by removing the Common Agricultural Policy subsidies to intensive industrial farming, plus internalising externalities, it would of course become economically rational to consume in such a way, encouraging more people to do so. These grassroots innovations may never supersede the supermarkets, but they remain an important demonstration of an alternative – very practical – vision, one which is essential for the achievement of a sustainable food system.

6
Sustainable Housing: Building a Greener Future

> People seem to change fundamentally when they gain the added security that comes from knowing they are capable of providing their own shelter. When a community of people posses that confidence and come together to help create one another's homes, it necessarily makes the world a better place to live (Steen *et al.*, 1994: xvi)

The bricks and mortar we live and work in are no longer keeping us safe as houses. Almost half the UK's carbon dioxide emissions come from heating and running commercial and residential buildings, and three million new homes are expected to be built by 2020. There is an urgent need to ensure new and existing homes are more sustainable in terms of both mitigating climate change (reducing carbon emissions), and adapting to the changing climate. There is no shortage of ideas – and practical demonstrations – about how this might be done, from high-tech smart houses which use the latest 'modern' construction methods and carefully monitor and adjust energy use in the home, to more 'down and dirty' low-tech solutions such as simple off-grid dwellings made of recycled consumer waste, and new social arrangements with shared neighbourhood facilities to promote social capital and cut resource use.

The examples of sustainable housing discussed here represents a range of low-impact, low-energy socio-technical systems, which although viable, celebrated and influential, remain marginal. The volume housebuilders responsible for providing the hundreds of thousands of new homes built each year have not adapted lessons from these green builders. In this chapter therefore, we first set out the policy context and historical development of sustainable housing in the UK, and then investigate the potential of some recent grass-roots innovations in the provision of housing to deliver more sustainable consumption. Finally, we examine the scope for these ideas and examples to spread from their green niches to influence wider mainstream practices.

The rationale for sustainable housing

The imperatives of climate change mean that our building technologies need to evolve to meet the demands of climate change predictions, while simultaneously reducing the contribution they make to CO_2 emissions. Housing plays a significant part in the UK's emissions profile (DCLG, 2007a). In 2005, 27% of the UK's CO_2 emissions (around 150 million tonnes a year) were attributed to heating, lighting and running domestic buildings; of this, almost three-quarters comes from space and water heating, and while appliances and lighting accounts for only around 22% of domestic emissions, current trends are for this to rise with new technologies such as digital radios, plasma TVs and air conditioning requiring higher energy inputs (DCLG, 2007a). In 2007 the UK's Department for Communities and Local Government published its blueprint for new housing over the coming 15 years entitled 'Homes for the Future: more affordable, more sustainable' (DCLG, 2007b), and an accompanying policy statement 'Building A Greener Future' (DCLG, 2007a). It identifies a growing housing shortage in the UK, caused by a combination of falling house-building rates, and rising numbers of households, many of them single-person households. The report sets out house-building targets of two million new homes by 2016 and a further million by 2020, but stipulates that homes must also become more energy-efficient to meet the government's Climate Change Bill targets for reducing CO_2 emissions to 60% of their 1990 levels by 2050. In the UK, while only around 1% of homes are built each year, by 2050 up to a third of the UK's homes will have been built since the present day, and 'we need a revolution in the way be build, design and power our homes' (DCLG, 2007b: 9). Therefore, in conjunction with its voluntary Code for Sustainable Homes (DCLG, 2008), the UK government aims to set progressively higher emissions-reduction targets through its building regulations, and so encourage improved standards in new-build housing, to achieve 'zero carbon' homes by 2016 (this is defined as zero net carbon emissions from all energy use in the home over a year, and applies at the level of the development, not the individual home, and at present does not permit offsetting to achieve the targets (DCLG, 2007a)).

There are social and economic, as well as environmental drivers for sustainable housing. Energy prices have risen dramatically in recent years, with average UK household gas bills rising by 109%, and electricity bills by 70%, between January 2003 and March 2008, with average annual household fuel bills amounting to £1060, resulting in a rise in fuel poverty. Energy-related indebtedness (measured in terms of consumers owing more than £600 on their utility bills) has risen sharply in line with these increases: between 2004 and 2007 it rose by 64% for electricity consumers, and by 19% for gas customers (Energywatch, 2008). At the same time, water supplies have been stressed in south-eastern England in particular (due to high population density, high levels of water use, increase in households and low rainfall), and across the UK water and sewerage prices have risen accordingly at above-inflation levels (see www.ofwat.gov.uk). Borrowing the language of carbon-neutrality, the UK government is implementing measures to promote 'water neutrality' in areas of new development to offset the water resource impacts of building new housing, with water conservation efforts such as rainwater harvesting, water conservation and metering. The aim is that the total water demand is unchanged after the development (Environment Agency, 2008).

Projections for the future indicate that these trends will worsen. Climate change is expected to bring more periods of extreme hot weather in summer, with peak summer temperatures up to 7°C higher by the 2080s than today, and the summer 2003 European heatwave when temperatures reached 38°C in the UK for the first time, would become the norm (Hulme *et al.*, 2002). Given these changing conditions, the buildings we live and work in may not be able to cope with extreme high temperatures in the summer. In the UK air conditioning is becoming normalised in workplaces, particularly in the south of England, to maintain thermal comfort, and their use is predicted to spread to domestic buildings over the next few decades. A recent modelling study found that in traditional 19th century terraced houses, and 1960s-built houses, the reduced need for heating over the next

80 years is offset by increased energy use for air-conditioning, resulting in overall increases in emissions of 30–40% by the 2080s (Hacker *et al.*, 2005). These calculations point to the need to retro-fit existing buildings, and design new ones in ways which do not rely on air conditioning to maintain thermal comfort, but rather draw on cooling socio-technologies traditionally employed in warmer climates, such as shading from the sun, thermal mass to stabilise temperature, passive heating and cooling systems, and afternoon siestas (*ibid.*; SDC, 2006).

Building a sustainable housing movement

While there is clearly an urgent need for new technologies and designs, it is also true that there are many technologies already in use, albeit on a small scale, which can deliver low- or zero-carbon homes, and some of these form the focus of this chapter. Many are descended from an earlier wave of sustainable housing activism and development, prompted by the 1970s environmental movement and the 'Limits To Growth' hypothesis (Meadows *et al.*, 1972), and later boosted by the push for greater energy efficiency and energy security prompted by the 1970s oil crises. Smith explains:

> The founding concerns of eco-house builders in the early 1970s were informed by the way existing house-building methods, technologies and services were wasteful of materials and energy, dependent upon finite sources for those materials, and highly polluting. The principle of 'autonomy' was developed in contrast to the incumbent regime (Smith, 2007: 436).

These eco-housing pioneers drew inspiration from Schumacher's 'Small Is Beautiful' (1993, first published in 1973) and his concept of 'appropriate technology', i.e. adopting a scale and complexity of technology appropriate to its setting, and similarly were concerned with the effects of housing on human health and spirit. To meet the demands of householders wishing greater self-sufficiency from expensive and potentially unreliable energy supplies, this meant generally low-tech solutions which could be self-managed to create 'natural' homes which 'support personal and planetary health' (Pearson, 1989: 12). A grassroots sustainable housing advocacy movement was formed, comprising activists, builders and academics, who shared many of the New Economics values of human-scale development, self-reliance, decentralisation and empowerment (Smith, 2007). Their activities tended to emphasise renewable material and energy sources, low-polluting materials, a concern with the overall lifetime impacts of the house (i.e. occupation impacts as well as construction impacts), and autonomy. In particular the experimental niche nature of much of this development lent itself to self-build by owner-occupiers, in contrast to the mainstream housing market where speculative mass-produced housing is the norm. The Centre for Alternative Technology was established in Wales in 1973 as a test-bed and showcase of renewable energy and appropriate technologies, community living and self-sufficiency. It later turned to public education and outreach as a means of spreading its ideas and lessons, and is still running extensive visitor programmes today, albeit in a quite altered mainstream context which now considers renewable energy as cutting-edge and desirable, rather than counter-cultural (see www.cat.org.uk).

During the 1970s much progress was made through experimentation and technological development for the duration of the energy crises; but once oil supplies returned to normal and the financial incentive for radical energy efficiency was removed, the movement largely lost the attention of mainstream builders and government. Nevertheless, development of sustainable housing continued during the 1980s and 90s. Vale and Vale (1991) define 'green architecture' as design which: conserves energy, works with climate, minimises use of new resources, respects its users and its site, and is holistic. Their proposals for an Autonomous House (Vale and Vale, 1977) were later realised, and the results showcased in an updated publication, the New Autonomous House (Vale and Vale, 2000). This was the first such self-sufficient residence built in the UK, purposely designed to look like a 'normal' house, it nevertheless demonstrated radical principles of self-sufficiency through energy generation, water harvesting and sewage treatment, and was not connected to mains utilities. Its design also capitalised on solar gain by using large south-facing windows to warm the house during daylight, and heavy ('thermally massive') walls and floors to store the heat and return it overnight, thereby significantly reducing the need to heat and cool the living space (known as passive heating and cooling). Vale and Vale subsequently worked on the Hockerton Housing

Project, a celebrated development of five autonomous earth-sheltered houses which have been found to use less than a quarter of the energy of a conventional house (Energy Saving Trust, 2003). Another leading light in this movement was the Findhorn Foundation's eco-village in Scotland, founded first as a spiritual centre in the 1960s, then developed as a demonstration site for green building and sustainable living from the mid-1980s onwards. With over 50 eco-buildings it now houses around 350 people, and continues to offer educational courses, while also providing a hub for the eco-village movement through founding the Global Ecovillage Network, holding major conferences in sustainable living, community and ecological design (Conrad, 1995; see also www.findhorn.org).

In addition to experimentation with building designs and energy systems, there has been a resurgence of interest in traditional building materials which could be locally-sourced from renewable or recycled materials such as straw-bale, wood, cob (mud and straw mixtures), reed and thatch, as well as alternative formulations of concrete using natural materials such as 'papercrete' (see Figure 6.1) and

Figure 6.1 Kelly Hart's papercrete home in Crestone, Colorado

'hempcrete' (see for example Steen *et al.*, 1994; Pearson, 1989; Hart, n.d.). Accompanying this has been a growth in social innovation such as housing cooperatives and co-housing (a type of community-based living where residents have their own homes and share some facilities such as laundry, a community hall and gardening), intentional communities and communes (see White (2002) for a recent overview of UK sustainable housing schemes in the UK, listing 81 exemplar projects ranging from low-energy single-household homes to large community self-build projects and eco-visitor centres).

With the advent of action to tackle climate change, eco-housing has become headline news again. In terms of the current policy agendas, sustainable housing is now primarily understood to mean 'energy-efficient' or 'low-carbon' housing. Lovell (2004) describes how the earlier 'advocacy coalition' of like-minded activists with shared deep-green values has been replaced by a 'discourse coalition' of actors with quite different perspectives on sustainable housing (e.g. eco-housing developers and mainstream housebuilders), but whose interests overlap on the topic of 'low-carbon' housing. For the latter, sustainable housing is about technology-intensive 'smart' housing which requires no change in householders' behaviour to deliver energy savings. The types of technologies employed might include movement-sensor lights, energy-efficient appliances and networked devices to 'intelligently' respond to residents actions (*ibid.*).

At the other end of the scale is 'low-impact' development, a term employed by grassroots builders concerned with minimising their ecological footprints, defined as development which:

> is temporary; is small-scale; is unobtrusive; is made from predominantly local materials; protects wildlife and enhances biodiversity; consumes a low level of non-renewable resources; generates little traffic; is used for a low-impact or sustainable purpose; is linked to a recognised positive environmental benefit. (Fairlie, 1996: 55).

Fairlie acknowledges that most buildings will not meet all these requirements, but argues that any truly low-environmental-impact development (as opposed to low-landscape-impact for instance) will conform to many of the criteria. Examples include temporary dwellings such as yurts, tipis and benders (tent-like structures made of bent and woven poles covered with canvas), as well as more permanent houses insulated with straw-bales and cob-built houses (*ibid.*). While these are extreme examples, they do demonstrate – as do all the grassroots innovations discussed here – that another way is possible, and that simple livelihoods can be obtained and managed in some unlikely locations – in woodlands, for example, or tending smallholdings, and without the need for high-capital investment property or high incomes to service mortgages and utility bills. A current exemplar of such development is Ben Law's wooden house in West Sussex, which was the subject of TV documentary Grand Designs. The house cost £25,000 and was self-built from chestnut and other materials obtained in the woodland in which it sits, self-sufficient for energy, water and waste, and a model marriage of low-impact design and lifestyle (Law, 2005; see also www.ben-law.co.uk). Another is Tony Wrench's off-grid wooden roundhouse built in the Pembrokeshire National Park in Wales (shown in Figure 6.2), built for £3000 from local natural materials and providing a sustainable rural farm livelihood, but the

Figure 6.2   Tony Wrench's low-impact roundhouse in Pembrokeshire, Wales

subject of a protracted planning battle (Wrench, 2001; see also www.thatroundhouse.info). This type of extremely low-impact housing tends to fall foul of planning regulations, both in terms of the construction of the dwellings, and the locations where people wish to build them. However, according to Simon Fairlie, a prominent writer in the field, this type of development is so sustainable – and has such a low ecological footprint – as to be sufficient justification for a new category of planning law, permitting low-impact development in rural areas to support sustainable livelihoods in the countryside.

Evaluating grassroots sustainable housing initiatives

Having reviewed the policy and research contexts of sustainable housing, attention now turns to practice. This section moves on from general discussions of sustainable housing to examine a series of practical initiatives in more depth. It draws on fieldwork (site visits and interviews) conducted in 2004 with grassroots pioneers of two different models of sustainable housing in the USA. Quotations are from personal interviews, unless otherwise referenced. These two initiatives can be considered emblematic actors in the sustainable housing movement. The intention in studying practice in the USA is to learn from pioneers whose ideas were developing ahead of UK experience (due to a range of cultural, regulatory and climatic factors which are discussed later on), and whose ideas were beginning to spread to the UK, to assess potential future opportunities and threats in the UK context. These initiatives all represent innovations for sustainable consumption that move beyond the technical, to examine the need for fundamental changes in values and behaviour, in developing eco-housing. The analysis concentrates as much on the individuals behind each movement as on the technical aspects of their building approaches, and uses the empirical, personal and contextual data obtained to make conjectures about the importance of social and cultural contexts in forming, developing and extending from green housing niches. Indeed, they are each shown to be deeply embedded in their social and cultural contexts, and the practical demonstrations they have achieved are tied intimately into the specific circumstances and lives of their advocates. In examining the impacts of these initiatives, we return to the five New Economics criteria of sustainable consumption developed in Chapter 3, namely the potential for localisation, reducing ecological footprints, community-building, collective action and building new infrastructures of housing provision, and the findings are summarised in Table 6.1.

Figure 6.3   Athena and Bill Steen (and son), Canelo Project, Arizona, USA

The Canelo Project

Leading proponents of the US straw-bale housing movement, Athena and Bill Steen founded the non-profit Canelo Project in 1990 to further their experimental and educational work (see Figure 6.3). Their seminal book 'The Straw Bale House' (Steen *et al.*, 1994) consolidated the then-emerging contemporary interest in straw-bale building among environmentalists keen to develop low-cost, energy-efficient buildings from natural, local materials, and inspired a growing movement in the US and around the world, with its practical advice, technical know-how and building plans. Further publications celebrate the beauty and diversity of applications of simple hand-built straw-and-clay construction and decoration techniques, and the vernacular of self-built shelter around the world (Steen and Steen, 2001; see also www.caneloproject.com). Canelo is located in southern Arizona, in the south-west United States, set among oak woodlands and high desert. The project aims to develop 'ways of living that connect us to others and the natural world' through 'an ongoing exploration of living, growing food and building that creates friendship, beauty and simplicity' (Canelo Project, n.d.). The 40-acre site holds the Steens' traditional adobe family home, an adobe guesthouse for visitors, and a dozen or so small straw-bale buildings used as accommodation and storage sheds, which have been constructed by participants of straw-bale construction and plastering residential workshops, and are evidence of evolving techniques and expertise. In addition, the Canelo Project works across the border in Mexico with local communities, teaching simple self-build techniques to enable groups of women to build each others' houses for around $500 each, and constructing a demonstration office building for NGO Save The Children in Cd. Obregon (*ibid.*).

These houses use bales of straw (an agricultural waste product) as large building bricks for the outer walls, which are then plastered with adobe (earth plasters). They are highly insulative, made entirely of local, natural, cheap materials, and are easy to work with, enabling wide participation in the building process. Straw-bale structures can be load-bearing (i.e. the roof sits directly on bale walls), or the bales can be used as in-fill between the props of a wooden-framed building. Often the earthen plasters of interior walls and floors are highly-polished, creating a waterproof, marble-like finish. Straw-bale can be used as simply a technical wall system in a mainstream house, offering greater thermal comfort and energy efficiency (see Figure 6.8), or it can represent a more sustainable way of living, suggesting a hand-sculpted, human-scale building, putting its owners more in touch with nature. The Canelo Project takes this latter approach, favouring the empowerment of small, cheap and simple self-build (see Figure 6.4), over the mainstream, impersonal and industrial-scale construction techniques. This type of building is well-suited to the desert climate, but simple adaptations to different climates can include termite/rodent barriers, stone foundations, and overhanging roofs to reduce rainfall on the outer walls (Steen *et al.*, 1994). Common concerns about strength, fire-safety and waterproofing of straw-bale buildings are generally unfounded, and plastered straw-bale buildings have been known to last 50–100 years.

The Canelo Project is clearly a pioneering grassroots initiative working towards sustainable consumption through research, experimentation, innovation and diffusion of lessons about straw-bale construction techniques. But as the following analysis of their sustainable consumption impacts shows, theirs is much more than a

Figure 6.4   The Pumphouse, a demonstration straw-bale shed at the Canelo Project

technical contribution. There is a strong ecological citizenship ratio-nale to this initiative.

To assess the Canelo Project's contribution to sustainable con-sumption, we use the five criteria developed in Chapter 3. First, the emphasis on using locally-available materials (mud and straw) for con-struction is a significant *localisation* impact of this type of building approach, quite distinct from mainstream building techniques. In addition, the Canelo Project's techniques and processes are adapted according to local cultures, materials and skills. For instance, in Mexico an earthen floor was associated with poverty, so the polished adobe floors were often concreted over. By emphasising the creativity and artistic techniques possible with earthen plasters, the material was given higher status and became more widely accepted. In addition, new recipe mixtures of mud and straw were developed to better suit the skills and tools available in different situations. Second, these building techniques imply a significantly *reduced ecological footprint*, principally through using natural, biodegradable carbon-neutral mat-erials, avoiding highly polluting materials such as cement, and then from providing greater insulation than conventional housing, so reducing energy requirements for heating and cooling throughout its life. In addition, the Project's ecological citizenship aims are to enable people to reduce their consumption levels and live simpler, more com-munity-oriented lifestyles, more in keeping with an equitable distribu-tion of resources. The rationale for straw-bale building was originally environmental – to reduce waste and energy use – and the Steens retain a keen awareness of the Canelo Project's role in enabling people to live equitably and comfortably within a 'fair share' of the world's resources – in other words, a much-reduced consumption level for the developed world. Bill Steen explains: 'when people get a take on green buildings in this country, we don't think beyond our borders – we don't look at the global picture – we don't stop to say "how much materials are there to go around?"'; he mentions that there are plenty of examples of high-consumption *unsustainable* straw-bale houses, which are 'totally unfair in terms of their size and the materials used'.

Third, the Canelo Project has powerful *community-building* impacts through its emphasis on low-cost, affordable materials and inclusive techniques. The hand-building technique using natural materials and little specialised labour lends itself to wider participation in building than is the norm when specialist skills and industrial tools and materials are used: 'People who might otherwise be excluded become directly and enthusiastically involved' (Steen *et al.*, 1994: 21). The Project's Mexican constructions were typified by women working barefoot and children playing around the site, mixing clays and mudding the bales, and communities coming together for bale wall-raisings over a day or two, with opportunities for social interaction and mutual support. From their perspective, straw-bale buildings can be sustainable, but they are not necessarily so, and it is the process of building, in relationship with nature, the materials, and with other people, that makes a building sustainable. In fact, as their work has progressed it has become the social and community aspects of straw-bale building which have become more prominent and valuable to the Steens. Fourth, and related to this point, the particular model promoted by the Canelo Project is one of *collective action* rather than an individualism. Athena Steen stated:

> There are other ways to build which are possible when you come together, than when you build by yourself. What's possible is *magic*. It's not always about the building, it's about building people, and communities, and relationships to nature. It was a vehicle for bringing people together, the building itself was a by-product.

In other words, the socio-technical innovation the Canelo Project promotes is structurally designed to enable collective effort. Finally, it is clear that the Project's approach to construction represents a fundamentally different system of housebuilding to the mainstream, and so develops *new infrastructures of housing provision*. It uses widely accessible and affordable techniques and materials, involves women and children in the building process, is suited to individual and community self-build, and aims to empower people to create their own low-cost environmentally-sound shelter.

Earthship Biotecture

Founded by maverick green architect Michael Reynolds (see Figure 6.5), Earthship Biotecture is dedicated to researching and experimenting with autonomous buildings made from waste materials, and spreading ideas about building zero-energy houses around the world. An Earthship is an 'independent vessel to sail on the seas of tomorrow' (Reynolds, 1990: 1), a building 'that will take care of you by interacting with and encountering the biology and physics of the earth' (Reynolds, 2004a; see also www.earthship.net). Earthships are thermally-massive buildings with large south-facing windows (in the northern hemisphere), designed to selectively capture solar energy (admitting sun in the winter and shading from sun in the summer) and store the heat in the mass. The thermal mass walls are constructed from old tyres rammed with earth, laid down like building bricks, and plastered to give smooth walls. Drinks cans and glass bottles are used as infill, and to create non-load-bearing internal walls, again used as building bricks and plastered over. With additional layers of insulation – earth banked up around the house at the rear and one or two layers of greenhouses at the front – an Earthship is like a cave, and the internal living space maintains a stable temperature and feels warm in winter and cool in summer, requiring no heating or cooling in any climate. Rainwater and snow is collected from the roof and filtered for drinking; greywater from washing is filtered through internal planters growing food in the greenhouses and stored for toilet-flushing and the garden, and blackwater sewage waste is treated in an external septic tank. These buildings generate their own electricity through wind or solar power, can be built cheaply using largely low-skilled labour, and are designed for lower-consumption lifestyles which empower their residents through utility self-sufficiency and its cheapness to run (Reynolds, 1990, 2000). The major objectives of Earthship Biotecture are:

- To reduce the economic and institutional barriers between people and sustainable housing;
- To begin reversing the overall negative effect that conventional housing has on the planet;
- To create a less stressful existence for people in an effort to reduce the stress that they in turn place on the planet and each other;
- To interface economics and ecology in a way that immediately and tangibly affects current pressing problems with existing lifestyles;
- To provide a direction for those who want to live in harmony with their environment;
- To empower individuals with the inarguable forces of nature;
- To find and distribute knowledge about sustainable lifestyles (Reynolds, 2004b).

In the early 1970s Reynolds originally began experimenting with autonomous passive solar houses built from waste materials as a response to unsustainable energy-intensive construction methods, housing shortages, waste management problems and the potential unreliability of energy and water infrastructure systems as a result of extreme climate conditions, political or economic collapse (Reynolds,

Figure 6.5   Mike Reynolds, founder of Earthship Biotecture, Taos, New Mexico

2004a). Climate change has pushed many of these issues even higher up the agenda, and Earthships can be seen as rational responses to the needs for climate change mitigation (through building which require no heating or cooling and are self-sufficient for energy, therefore having a zero carbon footprint) and adaptation to climate change (by providing resilient homes capable of maintaining stable internal temperatures, withstanding extreme weather events such as flooding, hurricanes, storms and heatwaves, and their effects on infrastructural energy, water and sewage systems) (Hewitt and Telfer, 2007).

Reynolds is based in Taos, northern New Mexico, at the Greater World Earthship Community, where self-sufficiency for thermal comfort and water are challenging tasks, with annual precipitation of around 300mm (approximately half that of London) and temperature extremes of –34°C in winter to 38°C in summer are not uncommon (Hewitt and Telfer, 2007). For three decades he has pioneered experimental buildings of this type, both with and without the support of local planning officials, and has incrementally improved the design and performance of his buildings as they have evolved, been adapted for other climates, and responded to commercial and regulatory pressures (see Figure 6.6). The Greater World Earthship Community is a 633-acre residential demonstration development of privately-owned

Figure 6.6 An autonomous Earthship, showing south-facing greenhouse front and earth-banked rear, Taos, New Mexico

Earthships in the high New Mexican desert, with approximately half of the planned 120 homes built, and over half the land retained as a communally-owned park. Buildings are available for private rental, offering visitors a taste of Earthship living, and educational seminars with hands-on building sessions further disseminate these ideas to wider communities. There are two further Earthship colonies nearby, built in particularly remote and inhospitable locations (e.g. on a mountainside) to demonstrate the viability of the concept (Reynolds, 2004a). Reynolds' team travels the world instigating Earthship projects, holding seminars and publicising their methods as a sustainable housing solutions. Although designed for the harsh, desert environment, Earthships have proved adaptable to other climates such as Jamaica, France and Japan. Two demonstration buildings have been built in the UK: the first a small visitor centre in Kinghorn, Scotland (see Kemp and Cowie, 2004), and the second a larger centre in Brighton (see Hewitt and Telfer, 2007).

The sustainable consumption impacts of Earthship Biotectures's initiatives can now be assessed. First, they enable much greater

Figure 6.7 Earthship walls made of tyres, earth, bottles and cans

localisation of construction, through the use of low-cost, low-tech waste materials (see Figure 6.7). Reynolds states 'I'm looking for a material that is indigenous to the entire planet. I've been all over the world, and everywhere there are tyres and cans and bottles' (Reynolds, 2004a). Of course, they also promote localisation of utility-provision, to the scale of the autonomous house self-provisioning for energy and water, food and waste-treatment. Next, they are fundamentally concerned with *reducing ecological – and specifically carbon – footprints*, through their independence from fossil fuels, their low-energy use design, and their use of recycled materials in construction. For example, the power system required to run an Earthship's essential services (pumps, lights, fridge) is very low, and can be run on 12v DC power straight from solar panels on site costing about $15,000. Inverters are required for supplying power to computers, TV, washing machines, and this energy should be used sparingly. In contrast, the solar panels required to meet a conventional home's power requirements would cost around $50,000, resulting in quite different price incentives and generally being too high to be widely considered. This illustrates how the Earthship principle of moderating consumption allows far greater scope for sustainable activities, while still offering reasonable access to modern luxuries: 'if you change what you need, in your head, then all of a sudden your life gets a lot easier' (Reynolds, 2004a). Recognising the need for global equity in resource use, Reynolds explains how self-provisioning is only sustainable if everyone else has the same standard of living, and takes an equitable approach to resource use and social justice.

Community-building is not an explicit objective of Earthship Biotecture's work, but it nevertheless does have some impacts in this area. First, it aims to enable low-income people to have secure, resilient shelter, and so a key element of its work is to provide plans and know-how for affordable, socially inclusive self-build. Second, while communities of replicated autonomous houses do not require collective infrastructure for their essential services (and so there are few *collective action* impacts), the initiative can be seen as an 'ultimate expression of personal empowerment and North American individualism' (Hewitt and Telfer, 2007: 114) but the Earthship colonies mitigate against this charge, to some extent.

Finally, and most fundamentally, the Earthship concept is a *new infrastructure of housing provision*, one which promotes resilience and

Table 6.1 Evaluating sustainable housing initiatives as a tool for sustainable consumption: key findings

Sustainable Consumption Indicator	Canelo Project	Earthship Biotecture
	Leading proponents of straw bale housing movement in the USA. Nonprofit organisation offering educational courses, demonstration projects and range of 'how-to' books. Favours self-build for the connection it offers between home and inhabitant. The process is at least as important as the outcome.	Research and demonstration of zero-carbon housing built with waste materials, requiring no heating or cooling, self-sufficient for energy and water, recycles greywater and treats sewage, and grows food.
Localisation	Construction materials are principally straw and mud, widely available and free in many parts of the world.	Majority of materials used are abundant and ubiquitous waste products from modern society – locally available and free. Building off-grid enables inhabitants to be self-sufficient for energy, water and sewage waste.
Reducing Ecological Footprint	Strong commitment to a globally equitable distribution of resources, hence advocates simple low-consumption living in the developed world. Very low ecological footprints of strawbale housing, both in construction and use through higher energy efficiency and thermal stability.	Major reductions in resource-use possible through self-sufficiency in energy and water, and through greater energy efficiency. High-consumption Earthships are possible, but not necessarily desirable; commitment to globally equitable resource use.

Table 6.1   Evaluating sustainable housing initiatives as a tool for sustainable consumption: key findings – *continued*

Sustainable Consumption Indicator		Canelo Project	Earthship Biotecture
Community-building	☞	Inclusive construction techniques, community-building (barn-raising), women and children involved too. Community-building is major impact of this type of housebuilding. 'Connecting people, culture and nature', building supportive networks.	Not explicitly concerned with community-building, but Earthship communities have been established.
Collective Action	☞	Small-scale activities, but with a strong sense of acting collectively such as through community-build projects, and empowering people within these groups.	Earthships are an expression of frontier individualism.
New Social Infrastructure	☞	Offers a system of housing self-provision which bypasses industrial construction techniques and technologies.	Allows householders to live independently of the utility grids, self-provisioning rather than relying on (potentially unreliable) infrastructure. Self-build, low-impact approach advocates home ownership without the need for high-income lifestyles to service utility bills and mortgages.

self-sufficiency, independence from mainstream modes of energy use and utility infrastructures. In his first book which sets out basic design and construction principles, he begins by stating:

> We need to evolve self-sufficient living units that *are* their own systems. These units must energise themselves, heat and cool themselves, grow food and deal with their own waste. The current concept of housing, in general, supported by massive centralised systems, is no longer appropriate, safe, or reliable (Reynolds, 1990: ii).

This new system of provision extends to social arrangements for financing housing too, emphasising low-cost self-build and the resulting low-income lifestyles that can be supported: 'It was such a freedom to not have a mortgage payment, to not have a utility bill, to know that no matter what happened to politics or the economy, I would have power, water, sewage, heating and cooling' (Reynolds, 2004a).

Diffusing the benefits of sustainable housing niches

> The process of building with bales includes the possibility of making a profound change in the fabric of human societies around the world. In fact, this vision is not exclusively a matter of straw bales; the questions we are trying to pose ... are basic: how do we build, and how does that process occur in relation to the community and to the life around us? Straw bales happen to be a material that has inspired many to look at the process of building in a different light (Steen *et al.*, 1994: xvi).

The two sustainable housing initiatives examined above represent a very particular, New Economics perspective on green buildings. Although they differ in terms of technical strategies and operational methods, their approaches have much in common. They are both bottom-up, value-led innovative organisations, founded by individuals following a social and environmental vision. They both practice and promote a new system of housing provision, embodying alternative socio-technical arrangements, to mainstream building practices in the developed world. We can see them as green sustainable housing niches, different in many dimensions from the mainstream, emerging from grassroots community activism, and aiming not only to thrive as niches, but also to influence the wider socio-technical regime.

In this section the potential for such diffusion of niche ideas is considered, with reference to the case studies described above. Smith describes the ways in which 'green niches are constructed in *opposition* to incumbent regimes. They are informed, initiated and designed in response to sustainability problems perceived in the regime' (Smith, 2007: 436), and they therefore have little compatibility with the main-stream system of provision. As a result, sustainable housing niches have little linking potential and growth prospects across all the socio-technical dimensions: guiding principles, technologies and infra-structure, industrial structure, user relations and markets, policy and regulations, knowledge base, and cultural meanings (*ibid.*: 429). This distinctiveness is evident with the sustainable housing examples dis-cussed here, and as the previous sections of this chapter have outlined, the characteristics of these niches which distinguish their system of provision from the mainstream are manifold. They advocate a small-scale, affordable, self-build approach to housebuilding; use local, nat-ural and recycled materials in inclusive and creative construction processes; they reduce consumption in building and inhabitation with highly energy-efficient designs and low-carbon technologies; they make explicit the consumption patterns and resource use which are otherwise inconspicuous, and challenge the accepted wisdom of cen-tralised power and water supply; and they embody ecological citizen-ship principles, offering a route to an alternative lifestyle: Athena Steen of the Canelo Project explains: 'People are becoming fed up with things the way they are, they're looking for a change. Somehow straw-bale holds that for people, it's a symbol of something different'.

Of course, these green housing niches do not exist in a vacuum; they have complex interactions with the mainstream regime, for better and for worse. Given the incompatibilities between the green niches and the regime, how have the Canelo Project and Earthship Biotecture fared in their efforts to spread their ideas and transform mainstream building practices? This diffusion of knowledge and practice can take three routes, namely replication at the same scale, upscaling, and translation of ideas to the mainstream regime; these are examined in turn.

The main way these two sustainable housing niches have grown to date has been through the replication of individual buildings, multiplying the base of green buildings at the same owner-builder scale, allowing for bespoke designs and adaptations as construction takes place alongside learning and experimentation. Concurring with the 'innovative niche' perspective of grassroots initiatives, Reynolds describes the approach Earthship Biotecture takes in terms of *innovation*:

> We are not a production outfit, we're an R&D [research and development] outfit, we're a demonstration outfit. Right now, we're going all over the world and planting seeds... and that's the best we can do because we don't have the government behind us, we don't have the corporations behind us.

He claims to have succeeded in developing these autonomous buildings only after being allowed the freedom to experiment and fail for 30 years, and is quite clear that 'the rules are inhibiting our evolution' (Reynolds, 2004a). Indeed, one of the main barriers facing green sustainable housing niches is posed by planning regulations and building standards which were not designed with these building methods in mind. In their study of autonomous sustainable housing, Van Vliet *et al.* found that 'new modes of provision can be limited by regulatory frameworks designed for public provision and infused with certain notions of what constitutes a safe and efficient method of supply' (Van Vliet *et al.*, 2005: 93).

Reynolds benefited from lax planning codes in northern New Mexico, and supportive local planning officers encouraging him to build experimental houses that would not be formally approved, but as political regimes have changed, stricter codes have been enforced and he has faced many legal battles to continue working to develop sustainable housing. At the same time the struggle to gain planning approval and meet regulatory codes spurs on greater innovation and improvement. In Taos, Earthship Biotecture has the world's first subdivision (residential neighbourhood) fully approved with no utilities, and they won the battle to build autonomous houses out of recycled waste, but they lost the battle to do it cheaply. High legal costs and investments to meet the regulations have resulted in land prices rising from $1600 a lot to $25,000, with implications for the inclusivity of the resulting community. His proposed solution is 'sustainable housing test sites' where experimental buildings could be constructed free of building regulations and the need for planning approval, to allow faster evolution of ideas and experience.

In addition to this development-within-the-niche, replication of both Earthship Biotecture and the Canelo Project's work has also occurred through publication of books and articles explaining their rationale and providing technical know-how for the self-builder, and also through educational courses offering hands-on experience with these unfamiliar building techniques. These methods have been successful in spreading ideas, best practice and lived experience among committed green builders and individuals searching for an alternative way of life. And this approach is slowly growing the movement across the world. In the UK, 'Amazon Nails' are a social enterprise working towards mainstream adoption of straw-bale building techniques, disseminating best practice and training community groups and construction professionals in low impact design and construction. They have been involved with over 50 projects in the UK, some with full planning permission and building regulations approval (others are used as sheds, animal shelters etc), and estimate that from the UK's first straw-bale building built in 1994, by 2001 the UK had approximately 70 such buildings in use (Amazon Nails, 2001). Although it is possible to obtain planning permission building approval for Earthships and straw-bale buildings in the UK (see Amazon Nails, 2001; Cowie and Kemp, 2004; Hewitt and Telfer, 2007), the unfamiliarity of local planning offices with the concepts make each application a laborious and potentially off-putting task for the green self-builder, and can prevent very low-impact buildings being developed at all in rural areas (see earlier discussion). Moreover, to the extent that this strategy relies on the continual recruitment of committed environmentalists, and ecological citizenship is a niche value, the scope to continue growing in this manner is limited in terms of numbers.

Compounding this limitation is the fact that mainstream framings of eco-housing 'continues to focus predominantly upon technical and economic aspects, whilst overlooking the social processes and guiding principles underpinning those developments' (Smith, 2007: 437). The specific circumstances which give rise to these green socio-technological niches relate to geography, climate, personality, economics, culture, politics and values: these socio-technical conditions cannot easily be replicated in an effort to reproduce innovative potential (Lovell, 2004; Shove, 1998).

Figure 6.8   A straw-bale house built in a mainstream style by Paul Koppana, Crestone, Colorado

The second route for niche sustainable housing practices to influence the wider housing regime is through scaling up the existing small-scale, one-off housing projects to industrial mass-production. This brings economies of scale to housebuilders, through standardisation of plans, materials and techniques, resulting in a profitable construction business. As a first step, the European branch of Earthship Biotecture have recently won planning permission to build a development of 16 autonomous Earthships in Brighton, England, delivering the first such buildings with residential planning permission in the UK (see www.earthship.co.uk), and the first Earthship colony outside New Mexico. Straw-bale building could likewise be adopted by mainstream housebuilders as an economically rational, energy efficient material (Amazon Nails, 2001). However, the models of sustainable housing discussed here are not necessarily practical for high-volume building. They both rely heavily on manual labour (making the techniques well-suited for self-build and community projects) which is costly for commercial businesses, and they were each developed in a context of cheaply available land and low density development, neither of which is applicable in the UK. Indeed, Hewitt and Telfer (2007) conclude that this combination of cost and the need for high density housing in the UK means that Earthships as they have evolved to date are impractical for a mass zero-carbon housing solution.

The third way that sustainable housing niches can influence the regime is to translate ideas and practices from one to the other, adapting them for the different socio-technical setting of the mainstream building industry. Examples might include the use of thermal mass to stabilise internal temperatures, shading from the sun, south-facing windows to capture solar gain, rainwater harvesting and grey-water recycling, microgeneration, etc. Modern methods of construction include using highly-insulated pre-fabricated wall panels, built in a factory and assembled on site; these can be filled with straw, hemp-crete and other recycled products, adapting niche material-use prac-tices to mass-production. For this to occur, a pre-existing condition of a crisis in the existing regime and an opportunity for niche prac-tices to inform mainstream solutions is required – this can be said to exist in the current need to develop low-carbon housing to mitigate climate change. The first step is for the mainstream conditions to open up opportunities for niche ideas to bridge the gap: government initia-tives to encourage greener building standards represents this type of top-down adaptation of the regime to adopt niche practices. But as Shove (1998) reveals, there is a chasm of meaning between the differing socio-technical contexts between niche and mainstream. Incremental improvements in insulation standards, for example, do nothing to challenge the mainstream paradigm of housebuilding reliant upon finite supplies of fossil fuels which niche autonomous housebuilders reject at the outset, and these conflicting perspectives prevent what might otherwise be seen as a purely technical transfer of knowledge. Therefore the regulation-driven mainstream only adapts in an ad hoc and piecemeal manner, failing to transform the regime (Smith, 2007). The second way to achieve a transfer from the niche is through the niche adapting itself to resemble the regime, as with intermediate pro-jects such as BedZED in London, a low-energy high density inner city development. This was a 'space where the practicability for volume housebuilders to operate more like green builders can be explored' and where 'values, processes and circumstances actually bring contrasting socio-technical contexts together' for more effective learning than is achieved simply through regulations (Smith, 2007: 439–40).

Building blocks and barriers to change

In this chapter we have examined the niche practices of two alternative, innovative sustainable housing projects, and considered their
potential to influence wider society. The initiatives were found to
successfully achieve sustainable consumption, according to the
New Economics indicators, through the use of local, recycled materials, inclusive construction techniques, and emphasis on reducing material throughput both in building and inhabitation of the
dwellings, and in enabling self-build and autonomy from mainstream utility services. They each present a viable – if currently small
scale – response to the need to mitigate climate change by reducing
energy use, and adapt to the demands of changing climates. However, despite both initiatives aiming to spread their ideas through a
variety of cognitive and social learning techniques, their influence
on volume housebuilding to date has been minimal. It is clear that
the innovations described here are radical versions of sustainable
housing, not necessarily suitable for scaling up or widespread replication; however, the scope for niche lessons to be adopted by mainstream builders is greater. Studying the interactions between niche
and regime reveals possibilities and obstacles for diffusion of ideas,
and this theme is returned to in Chapter 8.

7
Sustainable Currencies: Green Money from the Grassroots

> It would be very useful to have a local currency that rewards people when they co-operate just as automatically as conventional money rewards people when they set themselves up in competition with each other. If we could all experience the warmth that comes from belonging to a mutually supportive social network then not so many of us would feel the need to conspicuously consume unnecessary material goods.
>
> (Timebanks UK, 2007)

Could a new, 'green' money provide a solution to the problem of unsustainable consumption? A growing number of academics and activists around the world think it just might. The New Economics approach to sustainable consumption emphasises the social context within which consumption behaviour occurs, and is concerned not only with *what* is consumed, but also the *way* in which that consumption is variously shaped, enabled and constrained. This brings us to the question of money: what role does the money we earn and spend every day have in all this? Orthodox economists would argue 'very little' other than in terms of how much we have available; for them, money is a neutral tool, a lubricant in the economy which allows the exchange of goods and services to take place. Michael Linton, the originator of one of the most common types of local currency systems in use today (Local Exchange Trading Schemes, or LETS) disagrees. He claims that 'just as fish don't see water, economists don't see money' (Linton and Soutar, 1996). New Economists see money as a social mechanism, a facilitator of different types of relationships and behaviour, and they ask questions about how money could serve us – society and the environment – better.

From the coins in our pocket, to the reward points on our supermarket membership cards, the currencies we use offer distinct opportunities and constraints on our consumption patterns. Complementary currencies – parallel exchange systems using new types of money – have been widely advocated within the New Economics literature as a means of achieving sustainable consumption. Furthermore, they are claimed to articulate an alternative social infrastructure – a system of provision of money and exchange – which enables and incentivises particular types of social and economic relationships and consumption patterns. 'Complementary currency' is a generic term for the wealth of contemporary alternative exchange systems which exist alongside mainstream money. A wide range of complementary currencies have been springing up in developed and developing countries since the 1990s as a response to social, economic and environmental needs, in the form of skills-exchanges, modern-day barter, green versions of supermarket reward schemes, and even notes and coins.

This chapter reviews contemporary experience with complementary currencies as tools for achieving sustainable consumption. First the New Economics rationale for complementary currencies is set out, and mapped onto the five criteria of sustainable consumption derived in Chapter 3. Then three distinct types of complementary currency are described and evaluated according to the criteria outlined. Following this, the issue of how green niche innovations can influence wider regimes is addressed through an in-depth analysis of one of these initiatives in policy context, to uncover the potential and limitations it faces in diffusing sustainable consumption practices. Finally, the findings of this analysis suggest possible ways forward for community-based sustainable consumption, and these are discussed, together with appropriate policy recommendations.

The rationale for complementary currencies

This chapter is concerned with money as a type of socio-technical infrastructure, and efforts to create more sustainable alternative systems of provision to conventional cash. The key to understanding New Economics theories about the role and function of complementary currencies is to first ask 'what is money?' and 'what is wrong with mainstream money?' According to mainstream economic theory, money is a politically and socially neutral technology, with four core functions: as a medium of exchange, a unit of account, a store of value, and a standard of deferred payment (Lipsey and Harbury, 1992). According to this theory, the more mobile, efficient and widely accepted a currency is, the better it will perform its functions. Sociological and political, not to mention environmental critiques of this notion lead New Economists to challenge this assumption on a number of grounds (Hutchinson *et al.*, 2002; Dodd, 1994).

First, they argue that the functions of money – particularly medium of exchange and store of value – can conflict with each other. The fact that money is both a symbol (used for exchange) and a commodity itself (an item to be stored) encourages people to hoard money, removing it from circulation and thus reducing the amount available for transactions. It is a characteristic of modern economies that a shortage of money – supposedly the measuring stick of the economy – results in the paradox of having people with skills and labour to offer, plus work that needs to be done, but without the money to bring them together, the result is unmet needs and unemployed workers. This tendency has been observed by economists back to Gesell (1958) and Keynes (1973 [1936]), who promoted policies to ensure greater monetary circulation. A preferable solution, the new economists claim, would be to split the functions and have separate currencies for each purpose, so ensuring a ready supply of money for trade regardless of tendencies to save (store value) (Greco, 1994; Douthwaite, 1996).

Second, the mobility of money is not necessarily a good thing for local economies, according to these analysts. It results in 'capital flight' away from peripheral economic areas and towards centres, so draining regions and communities of the means of exchange. This centralising tendency, whereby money is concentrated in a few areas at the expense of other areas, is one of the economic costs of globalisation which the localisation movement seeks to address. National currencies are best suited to national-level and international transactions, and in performing this role, do not serve the needs of local communities well, according to the New Economics analysis, which criticises the 'dissociated' nature of modern money (Douthwaite, 1996; Robertson, 1999). Local economies are strengthened when money circulates many times within an area before leaving – known as the multiplier effect. New economics favours money that remains in a local area rather than migrating, and which is 'embedded' or founded within local social relations and environments, imbuing it with local significance and placing economic transactions and consumption itself within a profoundly social context (Greco, 1994; Lietaer, 2001).

Third, the current pricing regime upon which mainstream money is founded values some kinds of wealth and overlooks others, with profound implications for the signals sent by markets and hence development goals in general. Environmental and social costs and benefits are externalised from economic prices, and so are not accounted for in economic decision-making. This results in economic behaviour which degrades social quality of life and the environment, but which is entirely rational within the market framework (Jackson, 2004a). Economic rationality is a tightly-bounded world, divorced from ethical, social and environmental contexts, and arguably never intended to be considered away from these overarching – and fundamentally important – frameworks. Mainstream money is a tool used in a system which prioritises a narrowly defined range of economic activities. Its design and structure encourages users to value that which is scarce (and exploit that which has no monetary value). New economists – and increasingly they are joined by environmental economists working within the mainstream – argue that the dominance of markets at the expense of non-marketed aspects of life has gone too far, and argue for pricing to account for the full costs and benefits of activities, to enable genuinely rational decisions to be made which values all types of wealth, not merely that which is marketed (Robertson, 1999; Douthwaite, 1996; Daly and Cobb, 1990). This has implications for quality of life, justice, work and welfare.

Fourth, mainstream money and its system of exchange actively promotes particular types of behaviour and discourages others, and the implications of these effects are detrimental to sustainable consumption. For example, employment within the formal economy is rewarded while unpaid community labour is not; furthermore, the political structures surrounding the system of exchange reinforce this through the state benefits system by actively undermining people's capacity to undertake unpaid work and insisting that they enter formal employment. By redefining what is considered 'useful work' and 'wealth', New Economics aims to build a system of exchange provision which does not make these judgements, and which is more enabling of community participation and engagement through valuing all kinds of productive activity regardless of whether it takes place in formal employment or not. It suggests that the societal system of income distribution (currently based upon formal employment) should be altered, to remove the privileged position which formal employment currently has over other types of socially-useful work (Boyle, 1993, 2004; Robertson, 1999). Furthermore, despite claims that commodification is inevitably spreading and eliminating non-commodified exchange, there is evidence that non-market exchange (informal exchange networks and community currencies, recycling, second-hand goods, and so forth) is still a powerful force in industrialised economies (Williams, 2005). Consumers choose these alternative exchange networks for a variety of reasons, not only affordability, but also to experience and strengthen the anti-materialist values that such consumption embodies (Seyfang, 2001a, 2004c; Manno, 2002; Leyshon *et al.*, 2003).

This review illustrates how the current system of provision of money and exchange mitigates against actions and activities for sustainable consumption, and limits the scope of lifestyle changes which are possible within this system. The solution which a New Economics analysis suggests is to create new, alternative exchange systems which rectify these negative aspects; these are known as complementary currencies. The New Economics approach views all money systems as social infrastructure with in-built incentives, behaviour-framings and value. These can be structured to deliver sustainable consumption outcomes (Greco, 1994; Boyle, 2002; Seyfang, 2000; Lietaer, 2001; see Chapter 3 for a fuller discussion). For example Briceno and Stagl (2006) investigate complementary currencies as a type of 'produce service system', a socio-technical infrastructure whereby consumers access services, rather than owning products, for example through sharing and hiring goods. They find limited success at meeting physical subsistence needs, but considerable benefit in terms of social and psychological needs such as esteem, friendships, belonging and so on. Another example presented by Bob Swann (1981) of the Schumacher Society argues that specifically local currencies are tools designed at the appropriate scale for managing the sustainable development of self-reliant regions, and he suggests using energy – the kilowatt hour – as a universal standard of value; this measure could become increasingly pertinent with efforts to reduce CO_2 emissions to tackle climate change.

In addition to the 'social' currencies emerging across the world to tackle social, economic and environmental needs (see Chapter 3), a range of virtual currencies is now in use across the globe which are rarely thought of as alternative exchange systems, but which nevertheless function as mediums of exchange, units of account and stores of value. Air miles, for instance, are a virtual currency. They are earned when spending on everyday consumption goods and they can be spent on travel, so incentivising flying. Supermarket reward cards perform a similar function: they are given to consumers as rewards for purchasing at a particular supermarket (and so encourage loyal and increased consumption), and in turn can be spent in the stores or on special 'prize' items. Of course, these examples are corporate incentive schemes to encourage consumption, and so might even be seen as antithetical to sustainable consumption. But they serve to illustrate that complementary currencies are popular, in general use among the population, and are widely understood and accepted by the public (Boyle, 2003; Lietaer, 2001). Furthermore, as Air Miles claim to be a profitable *behaviour change* programme (Alliance Data, 2007, emphasis added), this raises the question of whether and how these tools can be adapted to encourage behaviour change towards more sustainable consumption.

Having described the problems associated with mainstream money and the conventional system of exchange, an alternative has been described: complementary exchange systems designed to address these problems and enable more sustainable consumption patterns. How effective are these complementary currencies at overcoming the drawbacks of mainstream money institutions, and facilitating sustainable consumption? The next section will review experience with three distinct types of complementary currency.

Evaluating grassroots complementary currency initiatives

This section reviews experience with three distinct types of complementary currency, each designed for a different purpose. It considers their characteristics and potential in terms of the five indicators of sustainable consumption defined in Chapter 3, namely localisation, reducing ecological footprints, community-building, collective action, and building new infrastructures of provision. The three complementary currencies discussed are: Local Exchange Trading Schemes (LETS) which aims to rebuild local economies; Time Banks which promote civic engagement and mutual self-help; and NU-Spaarpas, a 'green savings' currency which incentivises environmental lifestyle changes and sustainable consumption. Findings are summarised in Table 7.1.

Local Exchange Trading Schemes (LETS)

The most common type of complementary currency in the UK is LETS, Local Exchange Trading Schemes. This was developed in Canada, and introduced to the UK at The Other Economic Summit in 1985, out of which grew the New Economics Foundation. The first UK LETS was established in Norwich in 1986. A LETS operates a virtual currency to enable members to exchange goods and services without using cash, using local credits instead. LETS emerged in Canada as a response to the negative impacts of globalisation and economic restructuring, bringing unemployment and social fragmentation. This type of local money system was specifically designed to address the first two failings in mainstream money outlined above: namely that an abundant medium of exchange is required for a community to trade amongst itself, which circulates locally and cannot leave the area. LETS also seek to build community and create 'convivial' economies, embedded in local social relations. They aim to enable people to help themselves through work and exchange, without suffering externally-imposed limitations such as that of the systematic withdrawal of money (Lang, 1994; Croall, 1997).

Members of a LETS list their 'wants' and 'offers' in a local directory then contact each other and arrange their trades, recording credits and debits with the system accountant. The currencies often have locally-relevant, idiosyncratic names such as 'shells' in Kings Lynn, or 'bricks' in Brixton, and aim to instil a sense of local identity. No interest is charged or paid, so there is no incentive to hoard credits, and exchange becomes the primary objective. Most LETS are small, voluntary organisations run by local activists, but they have increasingly been championed (and sometimes funded) by local authorities under the aegis of Local Agenda 21 as a tool for local economic renewal, community-building and environmental sustainability. LETS has grown to about 300 schemes in operation at present, with an estimated 22,000 people involved and an annual turnover equivalent of £1.4 million (Williams *et al.*, 2001).

Figure 7.1   Nick's bicycle repairs on LETS encourage sustainable transport

The *localisation* impacts of LETS are evident in its design: this local money system was designed as a response to global restructuring, specifically to provide an abundant medium of exchange for a community to trade amongst itself, which circulates locally and cannot leave the area – so boosting the local multiplier (Douthwaite, 1996). Research has shown that LETS deliver small, but significant, economic benefits to members, providing new opportunities for informal employment and gaining skills, and enabling *localised* economic activity to take place that would not otherwise have occurred, and prompting some import-substitution (Williams *et al.*, 2001; Seyfang, 2001c). Some LETS have evolved to issue local currency notes, enabling the currency to spread further in the area – even through local businesses in some areas. There is some evidence that LETS can help people to *reduce their environmental footprint*. They promote local suppliers of food and other goods, reducing 'food miles' and the hidden costs of international transport associated with the conventional economy; they promote shared resources among members of a community, and so cut individual consumption, for example lift-sharing, hiring equipment and facilities; and they encourage recycling of goods, as members find a market for their unwanted items (Seyfang, 2001a). In these ways, LETS can be seen as a tool for building more self-reliant, socially-embedded local economies; indeed LETS has been championed as a tool for building green economies (Douthwaite, 1996).

The *social and community-building* impacts of LETS are very significant, as are intended to build community and create 'convivial' economies, embedded in local social relations. Where local notes are issued, they often affirm 'in each other we trust', 'in community we trust' (rather than 'in god we trust' as seen on US dollar notes). Research has found that they build social networks, generate friendships and boost personal confidence, in addition to being socially inclusive: they offer interest-free credit to financially excluded groups (Williams *et al.*, 2001; Seyfang, 2001c). Despite this strong community-building ethos, LETS is an individualistic tool, and does not presently appear to have any potential to influence *collective or institutional consumption*. Finally, LETS is constituted as a complementary money system, and attempts to *redefine the institutions* of exchange in the following ways: some LETS operate on a principle of increased wage equality (though this is not a requirement); the medium of exchange is abundant rather than scarce; and the money is locally bounded (North, 2006; Lee *et al.*, 2004).

However, despite this potential, LETS have remained small and marginal in economic terms, due to a number of internal and external factors limiting their growth: there are large 'skills gaps' which mean it is difficult to access staple goods and services through LETS; they tend to operate in 'green niches', attracting people who agree with the principle but have little time to participate, and indirectly excluding others; and government regulations deter benefit-recipients from participating by counting LETS earnings as equivalent to cash income and so potentially threatening means-tested benefits when levels exceed a given limit (Seyfang, 2001a, c; Williams *et al.*, 2001).

Time banks

The second wave of complementary currencies in the UK is 'time banks', which are based on the US time dollar model developed by Edgar Cahn, and aim to rebuild supportive community networks of reciprocal self-help, particularly in deprived neighbourhoods (Cahn and Rowe, 1998). 'Time banks' are a social economy innovation which reward participation in community activities or helping neighbours, and so aim to nurture social capital and networks of reciprocity. A time bank is a community-based organisation which brings people and local organisations together to help each other, utilising previously untapped resources and skills. It is a framework for giving and receiving services in exchange for time credits: each person's time is worth exactly the same – one hour equals one time credit, whatever the service given.[1] In this way, volunteer's hours are 'banked' and can be 'withdrawn' later when they need help themselves. A time broker manages the project and keeps a database of participants' needs and abilities. The types of help given are things like gardening, small DIY, giving lifts to the shops or hospital appointments, befriending, dog-walking, etc. These are things that family or friends might normally do for each other, but in the absence of supportive reciprocal networks, the time bank recreates those connections, and

---

[1]This is different to the high-profile BBC TimeBank media campaign which aims to attract people to traditional one-way volunteering through volunteer bureaux.

time credits are exchanged among participants as a form of time-based money or complementary currency.[2]

Time banks were invented in the mid-1980s by US civil rights lawyer Edgar Cahn as a response to the erosion of social networks and informal neighbourhood support which Cahn perceives as the bedrock of society (Cahn and Rowe, 1998). New Economist David Boyle had written about time dollars in the USA (Boyle, 1996), and invited Cahn to the UK in 1996 to spread the idea, and the first UK time bank was established in 1998 in Gloucester under the name Fair Shares. Time Banks UK, a national development organisation for time banking, was established in 2000 as a partnership between Fair Shares and the New Economics Foundation. In 2002 a national survey of Time Bank coordinators across the UK found that there were 36 active Time Banks with an average of 61 participants each (Seyfang and Smith, 2002). Since then, the idea has grown and by 2005 there were 70 active time banks across the UK with a further 70 being developed. This equates to an estimated 4000 participants, who have exchanged over 210,000 hours (Time Banks UK, 2005).

The stated principles of time banking are: recognising people as assets and that everyone has skills to share; redefining work to include the unpaid 'core economy' of work in the neighbourhood and community; nurturing reciprocity and exchange rather than dependency; growing social capital; encouraging learning and skills-sharing; involving people in decision-making (Cahn, 2000; Time Banks UK, 2001). Time banks are formal institutions which enable the non-profit-oriented exchange of non-monetised, services to meet social and economic needs, and which operate according to co-operative, egalitarian principles. Time banks are therefore 'alternative economic spaces' (Leyshon *et al.*, 2003), but in common with many other social economy initiatives, this alternative space is almost entirely dependent upon public (state) support, being reliant on grant funding. Indeed, Time Banks UK's aim is to promote the principles of co-production among mainstream public agencies, in order to meet the needs left unsatisfied by public spending cuts, help government meet its policy objectives for public services provision, and to improve public engagement with civic life.

---

[2]For more on the strategic development of complementary currencies in the UK and the learning from LETS to time banks, see Seyfang (2002).

The services provided on a time bank – neighbourly support such as dog-walking, gardening, small DIY tasks etc – tend to be locally-based by definition. But there is no net *localisation* effect, as the time bank creates new local networks and opportunities for exchange,

Figure 7.2 Angie offers permaculture gardening services on LETS

and does not substitute for imports. *Reducing environmental impact* is not necessarily a key aspect of time banking, but nevertheless it is being used to promote more sustainable consumption and environmental governance in a variety of ways. In north London for example, residents of an inner city estate will soon be able to earn time credits for recycling their household waste, and spend them on attending training courses or refurbished computers. Another London time bank rewards members with low-energy lightbulbs. Participation in groups which make local environmental decisions could also be rewarded. As indicated above, the primary rationale for time banking is *community-building*, and the projects are successful at developing social capital and new supportive networks. For instance, the Hexagon Housing Association in London covers 5000 properties in five boroughs (mainly the low-income areas of Southwark and Lewisham), and is incorporating time banking into its business model, as a tool to promote sustainable, cohesive communities. It is starting out by providing DIY and first aid training courses in exchange for time credits, and hopes this will empower residents to share skills, provide mutual support and develop a sense of community pride (and reduce maintenance costs for the association). Time banks attract members of the most socially-excluded groups in society (those who normally volunteer least), and are often introduced into marginalised areas where building trust and neighbourliness is a challenge which the conventional economy cannot meet. For instance, 58% of time bank participants have an annual household income of under £10,000 a year, compared to only 16% of traditional volunteers (Seyfang and Smith, 2002; Seyfang, 2003, 2004b, c). The benefits of time banking include increased self-esteem and confidence, gaining skills, growing social networks and building friendships, getting more involved in the community, and meeting needs – overcoming social exclusion and enabling active citizenship. For socially excluded individuals and communities, whose skills are accorded no value in the mainstream economy, the opportunity to be valued and rewarded for one's input into community activity and for helping neighbours, is enormously empowering.

There is also a *collective action* aspect to time banking. In addition to the 'community time bank' model, time banks can also be used as a 'co-production' tool to encourage people to become involved in the delivery of public services which require the active participation of service users in order to be successful, for example health, edu cation, waste management, local democracy, etc (Cahn, 2000). 'Co-production is a framework with the potential for institutions ... to achieve the elusive goal of fundamental and systemic change' (Burns, 2004; Burns and Smith, 2004), and the role and purpose of those institutions can be re-conceptualised from the bottom up, and reframed in terms of empowered participation and civic action. By rewarding and encouraging civic engagement, time banks could invigorate active citizenship. The 2004 national survey of UK time bank coordinators reveals that time banking is being used to pro-mote more sustainable consumption and environmental governance through 'ecological citizenly' action in a variety of ways. Mount Libanus time bank in Wales has organised a 'planning for real' com-munity visioning event, and plans to use the time bank to encour-age greater community involvement and sense of ownership of local environmental projects. This time bank and others reward participa-tion in local community forums and housing associations with credits, promoting engagement in local decision-making. Finally, the most significant benefit of time banking, for many participants, is the opportunity to redefine what is considered 'valuable', in other words: *creating new institutions* of wealth, value and work (Seyfang, 2004b, c). The radical of valuing all labour (or time) equally seeks to explicitly recognise and value the unpaid time that people spend maintaining their neighbourhoods and caring for others. Thus vol-untary work is rewarded and so incentivised (rather than squeezed out by the conventional economic system which accords it no value and so undermines social cohesion) thereby ensuring that vital socially reproductive work is valued and carried out (Seyfang, 2006). Time banks represent a new infrastructure of income distribution for society, where income is not dependent upon one's value to, and activity in the formal economy, but rather upon work – broadly defined – and this key issue is returned to later in this chapter (Seyfang, 2006).

Time banks aim to overcome the 'green niche' limitations of LETS by being based in mainstream institutions (health centres, schools, libraries), paying coordinators for development and support work, and most importantly, for brokering transactions between parti-cipants (Seyfang, 2002), but they still face obstacles in achieving their potential. These are: large 'skills gaps' in projects which again

Figure 7.3   Small DIY jobs are in demand on time banks

presents a limited range of services available; short-term funding
mitigates against projects which take a long time to become established (annual project costs were estimated to be £27,300 in 2002);
and reciprocity is slow to materialise due to cultural block, namely

the reluctance of participants to ask for help. In terms of government policy, while the unemployed are officially encouraged to participate in time banking (which is presented as mutual volunteering rather than an alternative monetary system), they may only exchange their credits for services, not goods. As time banks have developed in the US, material incentives for earning credits (donated refurbished computers, meals, household goods and so on) have been a major factor in encouraging participation from youth groups and the poor. This strategy is threatened in the UK by present government regulations which count their value as monetary income which is counted against social security benefits. Additionally, those in receipt of disability benefits face particular obstacles from the benefit system – this is discussed in more detail below (Seyfang, 2003, 2004b, c; Seyfang and Smith, 2002).

Nu Spaarpas

The Nu Spaarpas (NU) scheme is a 'green rewards' currency which has recently been piloted in the Netherlands and has been unresearched until now. This currency is designed to promote environmentally-friendly consumer behaviour, and operates like a reward card (Bibbings, 2004). Points are earned when residents separate their waste for recycling, use public transport, or shop locally. The points circulate in a closed-loop system, and card scanners in participating shops feed data into a central set of accounts. The initiative was a partnership between local government, local businesses, and non-governmental organisations – specifically Barataria, a sustainability consultancy organisation. NU was introduced in the city of Rotterdam in the Netherlands in May 2002, and by the pilot's end in October 2003, 10,000 households had the card, over 100 shops were participating, and 1.5 million points had been issued (van Sambeek and Kampers, 2004: 77).

It is an incentive scheme which specifically seeks to overcome the market disincentives to consume sustainable or ethical products, which are produced by the systematic externalisation of social and environmental costs and benefits from market prices. In other words, if mainstream money effectively incentivises unsustainable consumption, then NU is a prototype system which reverses those hidden subsidies by rewarding more sustainable behaviour. This is based on a marketing assumption that 'it is better to approach people in a positive and stimulating way than in a negative and restrictive manner' (van Sambeek and Kampers, 2004: 13) – in other words, promoting sustainable consumption using carrots rather than sticks (Holdsworth and Boyle, 2004).

Given that NU is a specific-purpose monetary tool designed to promote sustainable consumption, it is unsurprising that *localisation* and *reducing environmental footprints* are key outcomes of the initiative. As well as rewarding purchases from locally-owned businesses, extra points can be earned by purchasing 'green' or ethical' produce (such as organic food, fairly traded goods, recycled products, rental, repairs etc) at a range of participating local stores. The points are redeemed for discounts off more sustainable consumer goods, public transport passes, or cinema tickets (in other words, spare capacity in existing provision which incurs no additional costs), or donated to charity. Thus there are incentives to change consumption behaviour when both earning and spending the points, and private businesses benefit at the same time as public goals are met. However, in contrast with the other two cases examined here, there are no specific *community-building* impacts of NU: it is an individualistic mechanism, coming into play when individuals make consumption decisions (though it is socially inclusive, as points can be earned without financial expenditure). Despite this, it does have a role to play in channelling *collective action* through the public sector. NU was founded by Rotterdam Municipal Authority and prompted by several government objectives: reducing the volume of waste entering landfill, promoting public transport use, and generally raising environmental awareness and the practice of sustainable consumption. NU therefore has a direct impact on the provision of public transport, as well as waste separation facilities. Lastly, NU *creates new institutions* of exchange. If the market effectively incentivises unsustainable consumption (by externalising social and environmental costs), then NU is a prototype system which reverses those hidden subsidies by rewarding more sustainable behaviour, simply altering the relative prices of sustainable versus unsustainable goods. It anticipates the internalisation of social and environmental costs and sends appropriate price signals, and is easily understood by a public accustomed to savings points: 'the NU card scheme can present itself as a reliable channel for sustainability, and also offers low-threshold information that the consumer needs at time of purchase' (van Sambeek and Kampers, 2004: 77).

Table 7.1 Evaluating complementary currencies as a tool for sustainable consumption: key findings

Sustainable Consumption Indicator	LETS	Time Banks	NU
	Locally-based skill-swap scheme, aims to rebuild local economies. Usually run by community volunteers. Members advertise their 'offers' and 'wants' and trade in a virtual currency, with a central accountant keeping records. Offers interest-free credit.	Community-based reciprocal volunteering scheme, aims to strengthen social capital in deprived neighbourhoods. Usually run by charities and agencies. Everyone's time and skills valued equally. A broker matches participants to facilitate exchanges. Participation is more important than account-balancing.	Green loyalty points scheme to promote sustainable consumption, run by NGOs and local government. Points earned from sustainable consumption (recycling, buying local, buying organic etc), spendable on public services. Users have smart cards to swipe in stores.
Localisation	Economic tool, locally-bounded money to boost local multiplier, employment and self-reliance.	Community self-help is primarily locally-based anyway, so no net localisation.	Rewards buying from local businesses.
Reducing Ecological Footprint	Some evidence of reducing resource use: sharing facilities, recycling, localisation cuts transport costs (e.g. food miles).	Time banking concentrates on services, not material consumption. Some developments in rewarding recycling etc.	Incentivises recycling, public transport, local, organic and fair trade products and energy efficiency.

Table 7.1 Evaluating complementary currencies as a tool for sustainable consumption: key findings – *continued*

Sustainable Consumption Indicator	LETS	Time Banks	NU
Community-building	☞ Large social and community benefits, boosting social cohesion and inclusion.	☞ Very large social and community benefits: boosting social inclusion and social capital.	☞ Individualistic tool. But inclusive (not dependent upon spending money).
Collective Action	☞ Individualistic rather than collective action tool. Promoted by local government to mitigate poverty and unemployment.	☞ Promoted by central government to build capacity in voluntary sector and deliver public services. Could be basis for 'co-production' model of public service provision, and reward active citizenship.	☞ Individualistic tool, but promoted by local government. Influences public sector action in transport and waste.
New Social Institutions	☞ Some egalitarian measures e.g. minimising wage disparities. Capacity to value non-marketed work. Abundant medium of exchange. Localised monetary design.	☞ Central principle of valuing all types of work equally, rewarding unpaid community efforts. Reciprocity and mutuality.	☞ Points system adjusts relative prices to incentivise sustainable consumption. Anticipates internalisation of social and environmental costs and benefits.

Of the three alternative money systems examined here, NU is the most 'mainstream', as it exists comfortably alongside conventional money in regular everyday transactions. The pilot NU project adopted a high-profile, professional marketing approach to raising public awareness of the scheme, and cost €2 million to establish and run for the trial period; there are plans to make the card scheme self-sustaining financially, through charging clients (e.g. government) for meeting their objectives using the scheme (van Sambeek and Kampers, 2004). The main barriers to success faced during the project related to the experimental nature of the pilot, and to developing the project as it evolved – creating publicity material that successfully attracted participants, persuading retailers to take part and install the card scanners, etc – and as a result the organisers felt that making pronouncements about changes in consumer behaviour during the short pilot were premature. Nevertheless, they do point to a growth in the number of points issued from shops as time progressed, and a telephone interview with a sample of card-holders revealed that 5% of participants reported changing their behaviour (of buying organics, separating waste, and buying second-hand goods) as a result of the NU card – reporting that being 'rewarded' for making certain choices was the influencing factor. The NU scheme also made shop owners more aware of the different types of products they sold, but within the short lifetime of the pilot, no actual changes in provision can be attributed to the project.

Diffusing the benefits of complementary currency niches

As a social economy innovation, time banking is one with clear goals about influencing mainstream society and institutions – its founders deliberately oriented it away from the green enclaves which had constrained LETS, and framed it instead as a means of delivering public services. Given the UK's policy support for social economy initiatives as providers of public services, social inclusion, training and local governance (discussed in Chapter 4), how successful has time banking been at diffusing as a sustainable consumption innovation? The three routes for diffusing niche innovations are: replication, upscaling, and the adoption of ideas and practices by regimes, and this section examines the interface between the innovative niche, and the wider regime, to explore some of the dynamics of this interaction, and the scope for diffusion.

Initial indications are that diffusion should be successful, given official policy support of the time banking idea. Outlining the contours of the Third Way ideology which was highly influential to the UK's New Labour government, Giddens presents time banking as a model of innovative social entrepreneurship and governance by civil society, and argues that 'government should be prepared to contribute to such endeavours, as well as encourage other forms of bottom-up decision-making and local autonomy' (Giddens, 1998: 84). The task of building sustainable communities demands investment, from government and from local residents; and in both time and money, as Blair here asserts:

> As a nation we're rich in many things, but perhaps our greatest wealth lies in the talent, the character and the idealism of the millions of people who make their communities work. Everyone – however rich or poor – has time to give ... Let us give generously, in the two currencies of time and money.
>
> Prime Minister Tony Blair (2000)

Developing the capacity of deprived neighbourhoods to help themselves, and strengthening social capital, are key elements of the government's commitment to neighbourhood renewal and sustainable communities (HM Government, 2005). However, participation in volunteering has been declining in recent years, and new methods and tools are needed to encourage wider participation (Nash and Paxton, 2002). Time banks, as we have seen, are a direct response to these policy needs, and have been recognised as such in the Department of Health's green paper on Adult Social Care (which was publicly launched at London's Waterloo Time Bank) (DH, 2005), and in the Active Citizenship Centre's review of community engagement which highlights the achievements and potential of time banking in improving health (Rogers and Robinson, 2004).

With this official support, time banks have been relatively successful at the first path to diffusion, namely replication. A sizeable time banking movement has grown in the UK over just a few years, but they face a fundamental obstacle in growth by replication, which is the need for funding. In the 2002 survey, all the UK's time banks were found to be externally funded. Time banks do not rely on volunteers, but require financial support to pay the time broker's salary, for a publicly-accessible drop-in office, for marketing costs and so on, estimated to be £27,300 a year in 2002 (Seyfang and Smith, 2002). Funding for staff is crucial for time banks to successfully achieve their objectives of attracting socially excluded people in deprived neighbourhoods. Many UK time banks have been supported by grant funding from the National Lottery and various charities and trusts, and a funding cycle has been observed. Initially funding was more readily available for time banks, but over time it became harder to secure ongoing funding, or to increase the funding available for time banks overall. The consequence is that established projects close while new ones are begun elsewhere (*ibid.*).

Ironically, this situation has worsened because the second route to diffusion (upscaling) has not worked so well for time banking, for a number of internal and external reasons. Projects appear to work best when members all know each other, retaining a strong social bond. As schemes grow beyond a certain size (about 100 members), a greater distance enters the relationships between members, and the preference is to 'bud off' new time banks rather than continue to grow in size. So scaling up to really large projects is not a viable option for the current model of time banking which prioritises community cohesion (other models might not have this limitation). But there are other reasons why upscaling and attracting new participants has stalled. In order to promote the uptake of time banking among the poor and unemployed, in 2000 the UK government announced that time credits would not be counted as earnings, and so would not affect entitlement to income-related benefits. Neither are they counted as taxable income (Time Banks UK, 2006). This was a significant step for time banking within the UK: it overcame a well-documented policy barrier to participation in LETS; it ensured that the initiative had official support as a tool for tackling social exclusion; and the issue of participation affecting entitlement to state benefits could be dismissed. It was therefore framed as 'non-remunerative work', rather than 'economic activity'. However, the experience of time bank organisers and activists is that the benefits ruling does not go far enough, and there are three remaining regulatory obstacles to be overcome. First, the Department of Work and Pensions has stated that goods used as an incentive to participation on time banks (for example recycled computers which are awarded to participants for earning a certain number of credits), count as earned income (cited in Time Banks UK, 2006). In the USA, local businesses take part in time banks by donating surplus goods or services, which can be 'bought' for time credits. This is a useful way of attracting participants with economic needs, and widening the range of useful services that may be obtained on the time bank, and such a strategy in the UK would increase the benefits of time banking to the socially excluded enormously. Second, participants receiving incapacity benefits may find their payments cut because participation in time banks is presumed to demonstrate an ability to work (*ibid.*). Time bank organisers claim this is a mistaken and short-sighted assumption – the involvement of people with disabilities in community activities through time banking is first of all an effective form of occupational therapy, building confidence and skills, and second, only possible in many cases because of the high levels of support offered. Third, unemployed time bank participants – in common with anyone undertaking unpaid work in the community – find themselves pressured by current 'welfare to work' policy to enter the formal employment market, at the expense of their voluntary work (Seyfang and Smith, 2002; Burns *et al.*, 2005).

The third option of adapting niche lessons for application to mainstream regimes, is the area currently being pursued by time banking organisations to overcome the barriers of the other two routes to diffusion. Time banking's principle of creating reciprocal relationships based on equality, and valuing the time people invest in their communities, is a powerful one. Time banks could be incorporated into health, education and regeneration agencies, as well as charities and special interest organisation, as a tool to help them achieve their objectives. This adaptation of the time banking model to fit a mainstream regime would enable it to enrol thousands of participants, while relying on established public bodies to fund the project once incorporated into existing infrastructure. It could also be usefully adopted as a mechanism to boost public participation in local decision-making in areas with high levels of disenfranchisement. This could be both through official channels, e.g. Citizen's Panels or Social Inclusion Partnerships, or alternatively though community groups and lobbying organisations. In addition to time banks, other proposals have been put forward to achieve the same ends.

For instance Williams (2004) observes that unpaid community involvement is often required in regeneration partnerships, and that those who take part face the same lack of recognition for their efforts, and potential benefit penalties, as we have seen in the time banking example. He therefore puts forward two policy options to rectify this: first that the benefit recipients might seek to have their community efforts included under the 'voluntary and community' strand of the New Deal, in the same way as musicians have been recognised for their contribution to society and freed from the expectation of taking up paid employment, and second that 'active citizen tax credits' could be a system of rewarding voluntary work for the community. However, the lack of policy coherence particularly around state policy on benefits and work will remain a fundamental obstacle to the wider take-up of this niche practice. The causes and implications of this deep-rooted contradiction go to the very heart of UK public policy, and raise important issues about the potential of this niche practice to diffuse and influence the wider regime (Seyfang, 2006b).

The social economy has become a buzz-word in UK politics over the last decade as the 'third sector' or 'civil society' (that realm of economic and social activity between the public and private sectors) is increasingly seen as a source of social inclusion, cohesion, active citizenship, enterprise, training and employment, as well as public service provision and engagement in sustainable development practices (HM Treasury, 2002a, b; DEFRA, 2005a; DTI, 2002; Blunkett, 2003; Home Office, 1998; Giddens, 1998). However, the preceding review has hinted at contrasting theories of work, value and income distribution between UK public policy and elements of the social economy, as exemplified by time banking. In order to examine what is at stake here, these underlying values will be made explicit. The social contract embodied in the modern democratic welfare state holds that individuals who are able to work, have an obligation to do so and to thereby earn income to provide for themselves and their families; those unable to work are financially supported by the state. This contract is a powerful manifestation of the work ethic, and forms the basis of the system of income distribution in all modern economies: income entitlement is tied to formal employment and the unemployed are, by definition, socially excluded (Bauman, 2005). This system has been strengthened over recent years as the 'welfare to work' New Deal programme has emphasised even more the obligations of citizens to undertake paid work – even at the expense of commitments to childcare and community activities – and recent social inclusion policies have emphasised work as being the primary location for social inclusion (Byrne, 2005).

Yet this system of income distribution and its accompanying goal of 'full employment' – or the more modern 'employment opportunity for all' – is arguably partial in its scope and detrimental to cohesive, sustainable communities. It recognises only paid formal employment as 'work', so values only that work which has a value in the labour market, and stands in stark contrast to the active citizenship and civil renewal agendas discussed earlier. Indeed, participation in community and voluntary activities has been falling, and women (the traditional providers of unpaid community work) are doing less, as they are encouraged to undertake paid employment instead (Davis-Smith, 1998). In effect, this policy is stripmining communities of the very people they need the most – active citizens who work hard, on a voluntary basis, to meet social and economic needs in local communities – because they are officially viewed as being 'economically inactive' and are required to be financially self-reliant – i.e. not in receipt of state benefits (Burns *et al.*, 2005).

In contrast, time banking bucks the pricing and market system by giving a value – and incentive – to the work which is normally unvalued in society, yet which is essential for the development of sustainable communities. Cahn calls this the 'core economy' which *underpins* the public and private sectors – in other words is an essential prerequisite for a functioning society and economy. Social reproduction 'is the work that keeps local neighbourhoods safe, clean and inviting, keeps people healthy and happy, and enhances people's abilities as parents, friends, neighbours and potential employees – but never appears in government employment statistics' (Burns *et al.*, 2005: 3); it is quite literally unvalued in the conventional economy (Waring, 1988). As Bauman explains:

> Whenever one spoke of work, one did not have in mind household chores or the bringing up of children, both blatantly female provinces; but also more generally, one did not mean the myriads of social skills deployed, and the endless hours spent, in the

day-to-day running of ... the 'moral economy' (Bauman, 2005:
119)

Time banking aims to prevent this vital work from being squeezed
out by the pressures of the market economy, by building an alternative regime of work and income distribution which values and
rewards such efforts. One of time banking's primary attractions to
participants is its recognition and acknowledgement of the skills
and abilities of people who do not have a value in the labour market.
To use Marx's terms, it priorities 'use-value' over 'exchange-value'
(Amin *et al.*, 2002), and proposes an alternative system of societal
income distribution: one which is also based upon the work ethic,
but which redefines what we mean by work: i.e. it decouples income
from employment, and ties it instead to 'work' broadly defined to
include unpaid as well as paid exchange (Seyfang, 2003, 2004c). In this
way it speaks to the growing movement seeking to recognise and legit-
imise alternative forms of work organisation within modern economies (Gorz, 1999; Gibson-Graham, 1996; Williams, 2005; Robertson,
1985). As Lindsay attests:

If we focus on the value of the work rather than on perceptions
of what is economic based on narrow commercial definitions, we
see both the potential for expansion of the non-commercial
sector and the opportunity for participation in work as citizens in
a wide variety of contexts. We may work full-time or part-time,
paid or as volunteers in varying combinations in different stages
of our lives. (Lindsay, 2001: 119)

Indeed, the Third Way social democracy agenda holds that 'Work [i.e.
paid employment] has multiple benefits ... yet inclusion must stretch
well beyond work... An inclusive society must provide for the basic
needs of those who can't work, and must recognize the wider diversity
of goals life has to offer'. (Giddens, 1998: 110) If unpaid work in the
social economy is to be valued for its contribution to society – and the
active citizenship agenda suggests that it should – then government
must consider how it honours and incentivises that work. If social
economy initiatives are to grow and achieve their potential, then
this policy incoherence must be addressed, and efforts made to intro-
duce genuinely joined-up thinking around work, income and society.

Policy measures are needed which recognise – and reward – the valuable work performed in the social economy – valuable both to the individual and to society – and which thereby encourages participation in such activities by *all* groups in society.

Currencies of change

This chapter has investigated the scope and potential of three complementary currencies to deliver sustainable consumption. The findings indicate that each model of complementary currency successfully achieves some, but not all, of the criteria for sustainable consumption. This is due to the different purposes for which each currency was designed (i.e. whether there were primarily economic, social or environmental objectives). For example, LETS and time banks deliver large social and community benefits, but NU is focused instead on market transactions; meanwhile NU is specifically aimed at reducing environmental footprints through incentivising recycling and public transport use, while LETS and time banks only partially achieve environmental objectives as a by-product of other goals. However, there is one indicator of sustainable consumption which each of the complementary currencies delivers: they are all fledgling attempts to build new social and economic infrastructure founded upon New Economics values. They create new incentives, structures and institutions within which society transacts, so re-orienting it towards new sustainability goals. Indeed, they are prized channels for the expression of ecological citizenship values which are squeezed out of the conventional economy.

Furthermore, the three types of complementary currency are found to be complementary to each other: between them they succeed at achieving all the criteria for success, and so it might be argued that an effective sustainable consumption strategy requires a diverse range of alternative exchange mechanisms, each designed to target different areas of the development agenda. They demonstrate that the existence of plural monetary infrastructures is possible, and is effective at enabling more sustainable consumption patterns, albeit on a small scale. In so doing, they point to possible future developments which might take these principles and evolve them into something embedded within daily life for millions of people, translating these niche practices and lessons into mainstream practices and creating a web of interacting local currencies. These examples are suitable for local applications; other types of currency could similarly be designed for other scales of circulation and function, resulting a multi-tiered variety of specific-purpose currencies. For example, an international currency for global trade could co-exist with national currencies suited to taxation and public spending on infrastructure, sub-national regional currencies to promote economic development, and local economic and social currencies (Seyfang, 2000; Boyle, 2003; Robertson, 1999; Lietaer, 2001).

A number of policy changes are required in order for these grassroots currency initiatives to overcome the barriers they presently face. First, secure long-term funding is the greatest need identified to allow each of these projects to develop and grow over sustained periods, attracting broader cross-sections of members and becoming more established in society at large. Second, governments need to recognise the benefits delivered by participation in complementary currencies as being valuable for local economies, communities and environments. Complementary currencies benefit those on the margins of society – those on low incomes and outside the labour market for whatever reason; it is perverse that current state benefit regulations penalise those very groups from participating, and they need to be changed to reflect this. Third, government should embrace the possibilities offered by complementary currencies to deliver public services more effectively and achieve policy objectives across a range of areas – community capacity-building, poverty-alleviation, waste management, public transport provision, health and welfare – using alternative exchange systems as a tool to access places, social groups and motivations beyond the reach of the conventional economy. This examination of how one innovative niche for sustainable consumption interacts with the wider policy landscape and value regime has demonstrated that the diffusion potential for time banking is limited by an incompatibility between its expression of value and wealth, and that of the mainstream economic regime. Despite being framed as a public policy tool and gaining some ground in terms of incorporation into public service provision, time banking is ultimately an oppositional social institution, where niche practices constitute an alternative infrastructure of work, exchange, value and wealth creation.

By addressing these barriers preventing the diffusion of these niche ideas, government could do much to enable the spread of a social innovation with the potential to deliver more sustainable consumption.

8
Conclusions: Seedbeds for Sustainable Consumption

Sustainable consumption in all its guises – fair trade, local food, organics, energy efficiency, low carbon – is coming in from the cold. Green is the new black. It is starting to be considered fashionable, due to a concerted marketing effort to make sustainable lifestyles more desirable. Green consumers are increasingly called upon to lead the way in demonstrating social and environmental commitment through their purchases, sending powerful market signals to producers and retailers. But despite the headlines and high-profile marketing campaigns, this growing trend barely scratches the surface of the changes we need to make to developed country consumption patterns. There are significant problems with an approach which burdens individuals with the responsibility for achieving sustainable consumption, and which relies on conscious consumer decision-making, to the neglect of routine, habitual consumption. Barry Clavin of the UK's Co-operative Bank (one of the leading proponents of ethical consumption) celebrates the growing market for ethical consumption, but reminds us that ethical consumerism 'cannot be relied upon to deliver the significant 60–80% reductions in CO_2 needed' (in Co-operative Bank, 2007: 3). Indeed, it might simply be another reason to keep shopping and buying more products and continuing to fuel consumerist lifestyles:

> The middle classes rebrand their lives, congratulate themselves on going green, and carry on buying and flying as before. It is easy to picture a situation in which the whole world religiously

buys green products and its carbon emissions continue to soar (Monbiot, 2007: 27).

The imperatives of climate change, and the urgent need to make steep reductions in carbon dioxide emissions, are prompting critics to argue that current mainstream policy approaches to sustainable development are inadequate, and that a radical overhaul of developed country consumption patterns is required. This book has examined some of the proposed solutions to this crisis. It has examined grassroots innovations in three of the most fundamental systems of provision we encounter daily: the socio-technical systems which shape what we eat, where we live, and how we organise exchange between ourselves. At a more fundamental level, it has sought to provide evidence and critical appraisal of a 'New Economics' narrative about how we *should* live, for the long-term sustainability of our communities and planet.

The initiatives discussed in the previous chapters provide a small taste of what a New Economics sustainable consumption might mean in practice. Although apparently diverse in nature, they have much in common: they are all embedded within social economies, they utilise a range of organisational forms, marshalled to meet social and environmental needs, and they aim to establish new infrastructures of provision and social institutions based on ecological citizenship values. They all appear to overcome some of the obstacles faced by mainstream sustainable consumption strategies in enabling the practice of ecological citizenship, but still encounter barriers preventing them from achieving more widespread impacts. Drawing on innovation theory and the sustainability transitions literature, we can understand these barriers in terms of the clashes of oppositional systems encompassing practices, institutions, infrastructures and values. We can begin to conceive of the challenges ahead, if we are to overcome these barriers and harness the energy, ethics and innovative potential of the grassroots to catalyse wider societal change and achieve sustainable consumption. In this concluding chapter the main lessons from this approach are drawn out, to set out a new research and policy agenda for sustainable consumption.

The New Economics of sustainable consumption

The New Economics has emerged as an eclectic body of thought, incorporating insights from ecological, humanistic, institutional and behavioural economics to explain why modern developed economies are unsustainable, and to develop new ideas about how they may adopt more sustainable trajectories. Its four central themes – redefining wealth and progress, a broader conception of work, new uses of money and reintegrating ethics into economic life – comprise a new paradigm of socio-economic thought which foregrounds environmental sustainability and social equity (Boyle, 1993). Applying New Economics theories to the subject of sustainable consumption reveals much about the motivational forces driving current consumption patterns, and the scope for behaviour change as well as the imperative of wider transformations in social infrastructure. Max-Neef's theory of needs illustrates how consumption is directed towards the satisfaction of material and non-material needs, and importantly, how it is often mid-directed, failing to satisfy or triggering further social-psychological needs. Jackson (2004b) terms this process 'pathological consumerism', referring to a seemingly limitless desire to consume ever more goods and services. To the extent that mainstream sustainable consumption efforts are directed towards winning the consumer vote for greener purchases, they simply feed into this process; the New Economics offers instead a more radical analysis and proposes that sustainable consumption requires the development of five interlinked processes. These are: localisation, reducing ecological footprints, community-building, collective action, and building new social infrastructure or systems of provision. Tying these together is a new environmental ethic, Ecological Citizenship, which calls on citizens to take personal responsibility for the social and environmental impacts of their actions, but simultaneously to engage politically to transform wider societal conditions and institutions (Dobson, 2003). In many ways it overcomes the weaknesses of individualistic, mainstream approaches to sustainable consumption: Hassanein (2003) discusses the tension between individual quotidian political acts with regards to food consumption within the current systems of provision (reforms), and the large-scale collaborative action required to transform or recreate those systems (radical transformation). She concludes that a pragmatic democracy is needed to unite diverse actors and build coalitions among alternative food movements. Here, we can see that ecological citizenship bridges the divide between individual and collective action. It motivates private consumption choices, but at the same time speaks to a need for collective action to build new social infrastructure.

This theory of behaviour change founded on values is at the heart of the New Economics approach to sustainable consumption. It rejects mainstream market-based models of behaviour change, for focusing on materialistic incentives as goals, and for producing outcomes which are easily reversible when conditions change. Instead, it takes an approach modeled on political strategy rather than marketing, and frames behaviour in terms of collective activities and collaboration, rather than individualistic actions. This taps into people's enhanced sense of agency and environmental responsibility when motivated by intrinsic goals (fulfillment, community or intimacy) rather than extrinsic objectives (saving money, status display, self-image) (Crompton, 2008). Ecological citizenship fulfils this need, but requires spaces of practical expression for it to be realised, nurtured, spread, strengthened, and most importantly institutionalised and embedded within daily life. Though small in scale at present, initiatives which allow people to practise ecological citizenship values are important carriers of vision. The grassroots innovations presented in this book all offer that potential, allowing people the opportunity to co-create new social institutions based on their values, and framed in opposition to the unsustainable mainstream. For many participants, this largely symbolic outcome is the most significant and meaningful. With reference to comparable grassroots movements in the USA, Princen (2002: 41) remarks:

> From a production angle, the simple living, home power and local currencies movements are trivial instances of protest; they are of little political or economic consequence. From a consumption angle, however, they are concrete expressions of concern and resistance...[and] widespread discontent with consumerist society.

The presence of common values is important for grassroots innovations, not only in terms of uniting groups of people around particular objectives and practices, but also as a motivational factor behind the establishment of the niche activity in the first place (Lovell, 2004). Indeed, many green niches do not develop with diffusion in mind, but rather as spaces in and of themselves. These 'simple' niches are nevertheless vitally important as generators of ecological citizenship values and practices. Small-scale experimental activities are valuable demonstrations of alternative ways of working and living, and can inspire others to take action: 'their significance extends beyond their local contexts as they can provide glimpses of possible futures' (Georg, 1999: 465) and we can conceive of these activities as generators of ecological citizenship, as well as spaces for its expression.

The New Economics approach to sustainable consumption works with social institutions and contexts, recognising socio-psychological needs as consumption drivers, as well as questions of socio-technical infrastructure and systems of provision which effectively lock-in consumers to particular consumption practices, rendering them routine and habitual, and outside the scope of conscious consumer choice. This inconspicuous consumption – for example mains water provision, the electricity grid, private modes of transport – has implications for sustainability which are immune to exhortations for piecemeal, incremental change: they require system-wide transformation. The innovation and 'transitions' literature offers a useful perspective on this process, by modelling how processes of change (and agency) occur within this constrained landscape. It describes processes of transformation in societal regimes (institutions and infrastructure), and explores how transitions can be effected from current unsustainable practices, to more sustainable systems. One source of these transitions is innovative niches where change is seeded, incubating new techniques and social arrangements in order that they diffuse into wider society, and ultimately trigger wider societal regime change. Whereas this literature normally deals with technological innovation in the market economy, here the theories were applied to grassroots initiatives for sustainable consumption operating in the social economy. The case was made that community-based activities are a neglected source of innovation for sustainable development, and a conceptual bridge was made between two previously separate strands of theory and policy: technological innovation and community action (Seyfang and Smith, 2007). This theoretical approach was empirically tested through the critical examination of a range of case studies of grassroots innovations for sustainable consumption. The community-based initiatives in food, housing and complementary currencies were found to be generally effective at delivering sustainable consumption according to the New Economics criteria. Furthermore they appeared to satisfy multiple needs simultaneously (for example material food provisioning, belonging to a community, expressing beliefs and participation), offering lower-consumption routes to wellbeing. However, they all face barriers in diffusing their innovative potential to transform wider regimes. Green niche practices of this type are apparently successful at creating alternative infrastructures and systems of provision on a small scale, but struggle to translate these innovations into changes in mainstream systems.

The role of grassroots innovations in seeding change

The transitions literature describes three principal means for diffusing niche innovations into wider socio-technological regimes. These are: scaling up (niche activities grow in scale), replication (niche activities multiply), and translation (niche lessons are taken on by mainstream actors). The latter is the principal method for *changing* mainstream practices, as opposed to simply growing niche practices, and occurs most readily when the niche and regime resonate together in terms of values, organisational forms, contexts, metrics and so on. However, in the case of niche sustainable development activities, Smith's (2007) studies of green innovative niches illustrate the diffusion problems faced by alternative niches, due to their consciously oppositional framings. In other words, where grassroots innovations were established specifically to counter mainstream practices, as is the case with New Economics sustainable consumption niches, they are founded on quite different values, and aim to develop distinct sets of practices, to the mainstream. This dissonance with the regime makes it difficult to directly translate niche practices, but the distance can be reduced when the regime is searching for solutions to problems which place it under tension, or by either the niche becoming more like the mainstream, or vice versa. Using this translation process as a lens through which to view the experiences of the case studies in this book, offers a new perspective on the chasm between New Economics practices in the niche and mainstream values in the regime, and the potential for wider societal change. Reviewing the experiences of niche-regime interactions from Chapters 5, 6 and 7, reveals a set of cross-cutting issues which adds to the theoretical work presented in Chapter 4.

Replication

Grassroots innovations in food can develop as small-scale alternative systems of food provisioning to mainstream supermarket supply chains, such as the local organic food cooperative examined in Chapter 5. Successful replication of this type of initiative is evident, as similar projects (with local foci) spring up across the country, and is enabled by the existence of national networks of grassroots innovators in food, such as the Food Links organisations.

The cases of sustainable housing in Chapter 6 seek replication in terms of individual houses and builders. The two niches discussed both aim to disseminate their ideas and practices into wider society, building a movement of like-minded builders through books, videos, website information, giving talks, hosting visits, and developing best practice. They each publish informative guides to best practice and encourage replication of their ideas. But their emphasis differs. The Canelo Project found itself emblematic of an alternative way of life, attracting people who wished to find an alternative to consumerism and mainstream lifestyles, and enabling them to practice personal alternatives. To this extent, it remains a niche primarily focused on its intrinsic benefits. Earthship Biotecture takes a wider, more political approach, aiming to deliver extrinsic benefits with the explicit goal of transforming the way we think about building in the face of climate change. It does this by 'seeding' new Earthship projects around the world, training teams of people in Earthship-building techniques and spreading their skills and ideas internationally.

The experience of complementary currencies discussed in Chapter 7 illustrates further the problems and opportunities faced by innovative niches in influencing the mainstream. The currency initiatives examined here were all providing alternative means of exchange, in order to incentivise particular types of behaviour considered more sustainable than that encouraged by mainstream money. The movement has grown principally through replication of small-scale projects across the world, run by volunteer activists as expressions of resistance to the mainstream economic regime. In the UK, LETS appears to be in decline, while time banking has grown rapidly since its inception in the late 1990s, but now this movement faces a structural obstacle. Time banks require funding to maintain their infrastructure and brokering facility; this model is responsible for helping time banking succeed in the most deprived neighbourhoods, but it is also a source of restriction. Fixed amounts of funding for time bank projects means that replication is possible – but new projects are funded at the expense of old ones that cease operations – the overall number cannot continue to grow in this context.

Upscaling

Upscaling niche innovations to achieve greater economies of scale and participation is a major challenge. In the case of the grassroots food initiative studied, Eostre were concerned that their existing capacity might not meet additional demand, and they were not particularly keen to grow the niche. Customer concerns about poor presentation, food quality and inconvenience were not adequately addressed (risking customer loss) because it was assumed that shared ethos would be sufficient to overcome dissatisfaction with service and quality. Similarly, while their strongly-signalled deep green ethos would be a beacon for committed environmentalists, it might be off-putting to more mainstream customers. Overall then, Eostre appeared content within their niche and not interested in upscaling, but at the same time large-scale direct marketing customer-friendly franchises were moving into the area, threatening the small niche's existence.

The sustainable housing niches discussed in Chapter 6 are self-described experimental, continually learning and evolving initiatives, with neither the capacity nor the inclination to adopt industrial-scale construction techniques. Consequently, they do not address upscaling themselves, but they do advocate it, and suggest that their techniques would be suitable for widespread adoption. However, the techniques used (labour-intensive, using locally-available materials and requiring owner-builder involvement and commitment to work with their off-grid systems and so on) do not suit the mainstream-building regime, where standardised materials, plans and techniques are the norm. Similarly, the land-hungry, low-density housing models developed in areas where land is cheap and utilities are scarce once out in the countryside are not well suited for the UK without significant adaptation. Together, these factors make it unlikely that these innovations will be successfully upscaled in their present forms.

Complementary currencies face similar issues. Some (e.g. LETS) appear to be niche-bound, attracting primarily green activists and potentially deterring wider participation; others (e.g. Time Banks) deliberately try to overcome this barrier to growth using different framings, but they each face significant internal and external barriers to upscaling. Internally, if the range of services on offer is inadequate to meet members' basic needs, the schemes only appeal as fringe activities; there are also cultural barriers to overcome before people are comfortable interacting in this new socio-economic infrastructure. Furthermore, external barriers are equally restrictive. Despite their potential to deliver on several government policy active citizenship, volunteering and social care objectives, LETS and time banks are hampered because of fundamental differences in their socio-technical systems of provision. They offer an alternative means of distributing income (based on work rather than employment), and so are fundamentally in opposition to the regime. This results in clashes with related policy regimes, inhibiting participation by the groups who could most benefit.

Translation of ideas from niche to mainstream

As innovation theory suggests, grassroots initiatives can be a source of innovative ideas and experience, when tensions in the mainstream regime prompt a shift in practices, and a search for new ideas. The third type of diffusion occurs when niche lessons and practices are adopted by the regime, translated into mainstream contexts. The oppositional framing of green niches to the regime creates a wide gap between the two, resulting in greater difficulty translating ideas between contexts. To overcome this, there are two main options, described below.

The first route for grassroots innovations to become more easily translated into mainstream contexts is by the niche adapting its practices to become more like the regime, in an attempt to overcome the obstacle of widely different socio-technical systems. In the case of food studied here, the niche project was reluctant to be mainstreamed in case it meant losing their core ecological citizenship values and alienating their committed green customers. As a result the initiative appears to be niche-bound, as several internal barriers prevent them appealing to a wider customer base. However, there is a role for intermediary organisations here. For example, a smart organic supermarket in the area sells Eostre's produce in a manner more in keeping with mainstream supermarket retailing, and achieves higher standards of presentation, convenience and quality than the usual retail outlets associated with organic food, or the market stall.

Using intermediary actors to widen the customer base like this could be a successful strategy for niche initiatives who do not wish to adapt their core activities.

The Earthships case presents an interesting example of mutual adaptation bringing the niche and mainstream closer together to enable transfer of ideas and practices. After initially being given free reign to experiment and innovate, architect and designer Reynolds then spent many years battling against local planning officers to continue his work. Part of this struggle entailed adapting the niche to become more like the regime: by striving to improve the design of his buildings to meet building codes, and increasingly working alongside planning officers, even to the point of joining the local planning committee, Reynolds incorporated more mainstream practices and institutions into the niche. But alongside this negotiated compromise, the regime is now being changed from within, as Reynolds has a position of power within local planning infrastructure, and can demonstrate the viability of his projects, gaining greater influence over local development decisions, and working to establish the regulation-free 'test sites' he argues are necessary to allow rapid innovation (which would be another instance of regime governance adopting niche practices). In terms of achieving greater resonance between niche and mainstream, Reynolds has strategised effectively, making the most of an initial position of relative weakness, and doing so without support from above.

Complementary currency movements are facing a similar issue. Time banks in the UK have reached saturation in terms of available funding and the current model of (community-based) practice. The movement intends to evolve out of its niche by reframing itself and seeking to become integrated into existing public service provision, for example schools, hospitals, regeneration trusts etc. This approach avoids the need for funding for individual projects, and could potentially reach thousands of people at a time rather than dozens, but at the expense of the ties of support and friendship forged in close-knit community projects. The third group of complementary currencies examined in Chapter 7 (NU green reward points) appear far more mainstream than the community-based LETS and time banks, adopting the technology and language of supermarket loyalty schemes to appeal to a wide public. This grassroots innovative niche is deliberately designed to reduce the distance between itself and the regime, and offers perhaps the most promising model for diffusion. The more mainstreamed types of complementary currency (NU) managed to achieve far greater uptake than the niche-bound models (LETS and time banks), and they did so while retaining their original objective of reorienting price signals to favour sustainable consumption (a limited version of ecological citizenship). This key lesson is seen in other complementary currency initiatives, such as the Wedge local business affinity card (see www.wedgecard.co.uk), which is described in terms of boosting local independent businesses. Given this framing, it is perhaps unsurprising that all 194 of the UK's opposition Conservative Party MPs have joined (Wedge Card, 2008).

The second of these adaptations (where the niche takes on regime practices to enable more successful translation of ideas) was seen in the case of all the innovations considered here. In the case of food, supermarkets are increasingly competing for the newly-developed consumer market for local and organic foods, forcing the innovative niche to compete in terms that it would not normally prioritise: professionalism, presentation, efficiency, convenience. In the niche, customers' environmental commitment was considered motivation enough to overcome the drawbacks and inconveniences of an experimental emerging system of provision, but this limits the scope of the niche and inhibits its wider adoption. Mainstream supermarket encroachment into the niche market brings benefits and threats to grassroots innovators. On the one hand, it forces niche operators to raise their game to retain customers; but on the other, it undermines support for the alternative system of provision by attracting customers on the basis of tangible consumption goods (local and organic produce) but it provides these within the existing infrastructure, and does so at the expense of the intangible assets which the niche provided (community, connection with the countryside, supporting smaller growers, cooperative organisation). As a result, only the most tangible aspects of the niche practice are diffused to the regime, with a new focus on instrumental benefits rather than ecological citizenship; there is a significant loss of benefits from this adaptation. Another example of this adaptation was seen when public service catering was put under pressure in the regime following a popular television chef's criticisms of school catering. Local organic food was proposed as a superior alternative, and as the niche initiative had already made inroads into this area, it was held up as an exemplar of good practice and source of lessons as the regime sought to take on those practices.

Each of the eco-housing examples discussed here had experience with mainstream adaptations of their niche practices which they felt had lost much of their original value-driven purpose. They both took a low-impact, self-build approach to straw bale and Earthship-building, with a rationale of redefining lifestyles to reduce material throughput and empowering owner-builders. But both initiatives had witnessed their ideas being taken on board by people without such motivations, who intended to build high-specification, high-consumption buildings which just happened to use more sustainable building materials. In these cases, the conclusions drawn were that straw bale and Earthship techniques *can be* more sustainable than conventional building techniques, but they are not necessarily so. In other words, the material impact of the product is altered by its intended use; values are centrally important to achieving sustainable consumption. This supports Lovell's conclusions (2005) on eco-housing that as techniques are appropriated by the mainstream, devoid of their social contexts and unique processes, the factors which made them sustainable – and function – were removed, resulting in low take-up of the new technologies.

In the case of complementary currencies, there are some indications that the regime may be shifting to adopt more niche practices, so opening up a route for transfer of ideas. A recent report by the Commission on the Future of Volunteering (2008) in the UK made the important point that volunteers' efforts must be recognised and rewarded – not necessarily financially – and that unpaid work in the community must be valued alongside formal employment. If these principles were incorporated into policy, the regime would have made some major shifts towards the time bank niche, enabling far easier translation of the niche practices and ideas.

In addition, the lessons from experiments with complementary currencies can be applied to related initiatives at regime level. This can be seen in relation to current proposals to reduce carbon dioxide emissions by setting a national budget and issuing tradable carbon allowances (actually emission rights) to citizens (Fleming, 2005; Hillman, 2004). The allowance would decrease each year, so households would be encouraged to reduce their carbon emissions over time; proposals include holding carbon smart cards to be swiped at point of purchase alongside money, and so on. This system of personal carbon trading is effectively a new carbon currency. It operates as a medium of exchange (permits are surrendered in exchange for the CO_2 emissions associated with purchased goods and services – petrol, electricity, heating oil, flights etc); it is a unit of account (representing permission to emit a standard unit of CO_2), but it is not a store of value (permits expire after a certain time). Although carbon credits can be exchanged for money, they are nevertheless spendable in their own right, and can be considered a 'limited purpose' or 'special money' with particular distinguishing sociotechnological meanings which will influence its use (Dodd, 1994; Zelizer, 1994). Indeed, internalising carbon emissions into decision-making, and making them tangible, requires that consumers begin to count the carbon cost of their actions. Carbon allowances would be conceptualised and used ('spent' and 'saved') much as other virtual currencies (e.g. air miles) are at present, and it is useful to see PCT in this light to consider how public experience with using CCs offers lessons for PCT, despite vastly different scope, scale and development (Seyfang *et al.*, paper submitted to Ecological Economics).

Having reviewed the processes of dynamic interaction between grassroots innovative niches and mainstream regimes, it is clear that while significant obstacles remain, there are possibilities for translating lessons and ideas from niche to regime, and so diffusing new ideas about sustainable consumption into wider society. However, it is also evident that this process involves adaptation, reformulating niche knowledge and practice to sit more comfortably within the mainstream, and consequently blunting its oppositional edge. It is furthermore clear that unequal power relationships exist between niche and regime, and this imbalance influences the ways that niche lessons are adapted, and the leverage that grassroots innovations have – or do not have – in ensuring their ideas are adapted as intended (Smith, 2007).

Harnessing grassroots innovations for sustainable consumption: lessons for policy and research

Several key lessons for policy and research emerge from the present study, concerning the ways in which innovative niche practices connect with mainstream activities, and the potential for these grassroots

Table 8.1 Diffusion of grassroots innovations from niche to regime

Diffusion Route	Sustainable Food	Sustainable Housing	Sustainable Currencies
Replication	Growth in local organic food initiatives, farmers' markets. Local sustainability focus encourages replication of local initiatives rather than emulation of regime national/international food systems.	Niche actors promote replication through informative guides, inspirational talks and videos, experiential hands-on workshops, and through international seminars/training sessions. Growth of self-build and bespoke eco-housing limited to niche of committed green activists. Planning restrictions inhibit wider take-up.	Complementary currencies are set up at a variety of scales. Community-based projects workmore successfully at relatively small scale, so 'bud off'when they get big – resulting in many small systems. Limits to funding prevent the overall number of projects growing. Voluntary initiatives suffer burn-out.
Upscaling	Initiatives can grow to achieve economies of scale (e.g. through franchising) but need to attract wider range of customers (extending from the green niche), and also recruit more producers to meet additional demand.	Self-build eco-housing model is labour-intensive and best-suited for experimental and participative construction – not appropriate for industrial-scale volume housebuilding. High housing density and land costs in the UK also mitigate against techniques developed where land is cheap and abundant.	Community projects work best at small scales. Growing projects means attracting wider range of participants , overcoming internal obstacles (skills gaps, cultural barriers) and external limitations (funding and benefits policies)

Table 8.1 Diffusion of grassroots innovations from niche to regime – *contiued*

Diffusion Route	Sustainable Food	Sustainable Housing	Sustainable Currencies
Translation to mainstream context (regime moves closer to niche)	Supermarkets promote locally produced and organic food, competing on grounds of convenience, presentation, and quality, but cannot deliver the social and ethical benefits of the niche. Public pressure forces public sector to look to niche for lessons in healthier schools catering.	Rising oil prices and regulation to mitigate climate change put pressure on regime to find sustainable housing solutions. Regime takes on some niche practices, framed as 'low-carbon housing', but does not engage with social and cultural values of the niche. Niche actors are enrolled within the regime, aiming to transform it from the inside.	Initiatives to recognise and rewardvoluntary community efforts are adopting time banking practices. But regime system of income distribution fundamentally opposes niche model, limiting scope for integration.
Translation to mainstream context (niche moves closer to regime)	Niche may not wish to adapt for fear of alienating core customers, but intermediary actors (e.g. small green supermarkets supplying niche products) can bridge the gap.	Niche responds to regulatory pressures and strives to meet regime standards, to survive. Succeeding in some areas (e.g. meeting building regulations) can mean failing in others points (e.g. affordability, accessibility).	Complementary currency efforts to reframe in terms of public service provision, seeking integration into regime. Mainstream green loyalty system mirrors existing rcustomer eward schemes to attract wide participation, but lack community-building benefits.

seeds of change to germinate and take root. The first of these requires researchers and policymakers alike to recognise the contribution that grassroots initiatives can make to meeting policy goals, and this relates to metrics. The New Economics evaluation framework used in this study offers enormous potential for further research and refinement, to aid the understanding and development of new tools for sustainable consumption. The identification of a set of indicators has highlighted precisely how some initiatives score better than others, in different areas, and allows policymakers to work with a simple checklist of factors to consider. In particular, a set of more detailed examinations of initiatives will facilitate the wider adoption of the evaluation tool in assessing progress towards sustainable consumption. Furthermore, this appraisal framework allows for a richer examination of the needs-satisfaction impacts of consumption initiatives, enabling social, cultural, ethical as well as economic and material needs satisfaction to be evaluated. As such, it affords an insight into the types of initiatives which satisfy multiple needs simultaneously, so reorienting consumption patterns away from pseudo-satisfiers, and reducing material consumption.

Building on this, and developing the theoretical framework proposed in Chapter 4, it is vital that community action is recognised as a previously neglected site of innovation for sustainable development. Harnessing grassroots innovation requires management akin to that common with technological innovation. The innovation literature uses the term 'strategic niche management'; to describe how policy measures can be used to nurture niches until they are institutionally embedded, and have the capacity to expand into wider society. Considerable work is needed to transfer – and adapt – knowledge and lessons from the innovation literature – which traditionally concentrates on technological innovation in a business setting – to this new context. There is an urgent need to apply such strategic thinking to the management of grassroots innovative niches, to identify the conditions required for such niches to develop and grow, and to ensure that policy and funding support is in place to let them expand, become institutionalised, and transfer their practices to the mainstream.

Two recent reports shed light on the challenges faced when making this contextual shift. The Young Foundation's 'Social Silicon Valleys' (Mulgan *et al.*, 2006) proposes a set of measures to mobilise resources, expertise and enthusiasm to accelerate social innovation. These include new funding sources, special innovation incubators (akin to business incubators, using a 'protected niche' approach to development), new institutional bodies and support networks, and new ways of rewarding social innovators not motivated by profit. Similarly, NESTA (the National Endowment for Science, Technology and the Arts) aims to model how social innovation occurs, and identify the factors critical for its growth, and for its ultimate aim: 'changing how societies think' (Mulgan *et al.*, 2007: 22). By applying traditional innovation theories to this field, it identifies a series of 'innovation system' gaps where social innovators rely on ad hoc, voluntary efforts for coordination, whereas business innovators have well-developed processes and procedures, funding and support. Both reports highlight the paucity of research in this area, and this book identifies an even more pressing need where the subject is socio-technical innovation in the social economy, where specific innovation occurs as a result of combining particular non-generalisable actors and processes.

Third, it is evident that top-down government support is essential in enabling bottom-up innovations to flourish and thrive – these nascent initiatives cannot be expected to deliver societal change unaided. While grassroots initiatives have been championed here as innovative spaces where sustainable consumption can be practiced, the discussion above illustrates how limited their diffusion potential is, in the absence of enabling policy frameworks (see also Smith, 2007). Official policy commitments to sustainable innovation and community action provide rhetorical resources at least, but state support for grassroots innovations must become tangible and substantive. Governments can be pro-active in funding grassroots innovations and providing space for their development and growth. It is essential that existing policies do not undermine their ability to develop. This point is crucial: alternative initiatives for sustainable consumption do not require top-down government *control*, but rather the ability to grow externally to the mainstream without being squeezed out of existence by a policy-making process which is blind to their contribution to sustainable consumption and ecological citizenship. While the conditions that lead grassroots innovations to emerge may be locally-specific, and dependent upon particular types of people, it is impossible to simply legislate for greater grassroots innovation to occur. But what is achievable is to provide the right conditions for them to thrive, grow, spread ideas and develop networks, so encouraging them to thrive as they emerge naturally.

These conditions include not only financial support for grassroots innovators, but also for the institutional bodies and activities that would help overcome the structural weaknesses of community action in relation to technological innovation. For example, intermediary organisations can help to bridge the gap between grassroots and regime, speaking the language of both to help transfer ideas beyond the niche, and could be supported to develop in specific targeted innovative areas. In addition, networking organisations can capture learning and institutionalise it, sharing resources among partner groups, and reducing the entry barriers to participation facing potential new niches. The Global Ecovillage Network is one such organisation, which holds international conferences, distributes a newsletter, and promotes knowledge-sharing among member groups, all of which in turn help to strengthen and grow the movement (Conrad, 1995). Such bodies deserve to be recognised as new institution-builders, responsible for nurturing innovation for sustainable development, and financially supported accordingly.

By their nature, most innovations fail; the same is true for sociotechnical innovations, yet current institutional systems of funding tend to mitigate against risk-taking at the grassroots. Policy changes to allow these initiatives the space to experiment and fail, free from regulatory controls which hamper evolution of ideas, would encourage a much speedier rate of innovation. As already mentioned, community action incubators could be one such possible way forward, hot-housing grassroots socio-technical innovations in a supported environment. This could take several forms. For the more technologically-oriented innovations, it might consist of 'test sites' of the type proposed by Earthship housing pioneer Mike Reynolds. These would be physical spaces where normal rules of planning permission and building standards do not apply, to allow innovators to develop their ideas more quickly and freely, learning-by-doing, and sharing ideas and expertise. For social innovations, the equivalent might comprise supported efforts to cluster grassroots innovations and so benefit from the synergy, scale and 'normalisation' that comes when there are multiple overlapping green initiatives operating together. An example of this might be the rapidly growing Transition Towns initiatives, bringing together a wide range of local environmental and social actions under the umbrella of mitigating climate change and making the transition to an oil-free future (see www.transitiontowns.org). These clusters aim to develop a critical mass of participation, actions and values around behaviour change, so reaching a 'tipping point' and achieving wider institutionalisation of greener lifestyles. A final possibility for incubating grassroots socio-technical innovations takes a different approach to introducing people to the possibility of normalised sustainable consumption practices. Rather than changing the social context of a town, another option is to temporarily offer people the experience of living in a (small scale) world where ecological citizenship is the norm. One example of this is Dance Camp East, a community arts summer camp which aims to build sustainable communities. It offers a temporary physical and social space where 500 people gather and explore new ideas about how to live, and the alternatives that are possible, living simply and in community. Perhaps its most powerful impact is in offering a window into another world where sustainable development is the norm, and socio-technical institutions reflect that basis. This insight is potentially transformational – on returning home, participants view their 'normal' life and the infrastructures underpinning it (mains water, electric lights, globally-transported food) in a new light of possibilities for change (Seyfang *et al.*, 2005).

However, as Smith and I have argued elsewhere, if an innovation agenda is brought to the grassroots, a number of governance issues are raised for both policy and research (see Seyfang and Smith, 2007). Grassroots innovations will become boundary objects, interpreted differently by networks of actors encountering one another's interests and commitments around the niche. Government departments have their own objectives; technology developers have a different modus operandi to grassroots idealists; eco-entrepreneurs seek commercialisation, moving innovations from social economy to market economy; and academics bring their own agendas. Through niche engagement, and associated social learning, the positions and commitments of some actors will alter. The need for research into the contexts, actors and processes under which niche lessons are able or unable to translate into mainstream situations (and transform sustainabilities) will become even more pronounced (Smith, 2007). This raises important issues in research ethics, since it is vital to be respectful of the grassroots agenda, ensuring the intrinsic benefits of grassroots innovative niches are not undermined by diffusion interventions. Seeing the grassroots solely as business incubators would denude them of important and diverse features. The wider diffusion of niche elements through the market can be a welcome contribution to wider (shallower) greening. But other, less immediately commercial elements of grassroots niches remain potential sources of strategic diversity, important for living with the uncertainties associated with sustainable development (Stirling, 1998).

Policy and research into grassroots innovations must nurture mutually beneficial relationships with niche activists. The emerging agenda should consider how best to reward and encourage innovative behaviour at the grassroots – given that rent-seeking behaviour is not the primary motivation. Fundamentally, this is a question of how one traverses the interface between the social and market economies. A twin track approach is needed. On the one hand, research and policy that contributes to the creation of diverse grassroots innovations and engenders a variety of sustainable practices is needed. On the other, research and policy is needed that learns from this wealth of alternative means of provision and embeds that social learning into the mainstream. Policy measures must put the incumbent socio-technical regime under tension and prompt wider searches for (grassroots) sustainability innovations, for instance by adopting new measures of wellbeing and progress which challenge the doctrine of continued economic growth, or by supporting other social institutions such as churches and schools in encouraging citizens to question consumerism and materialistic lifestyles. Researchers can contribute by bringing a reciprocal learning approach to grassroots innovations, e.g., through action research (Stringer, 1996). Engaged researchers can offer services such as evaluations and policy analyses which grassroots initiatives themselves may lack the capacity to produce, but done in a way that challenges conventional analytical criteria. This could prove an essential strategic response to the ethical dilemmas noted above. Existing understandings of community action in sustainable development need reconsidering through the lens of grassroots innovation. Survey research can map the extent, characteristics, impacts and outcomes of grassroots innovations. In-depth qualitative analysis is needed to understand conditions for the germination of innovative processes at the grassroots, and the conditions for successful diffusion, examining the role of social networks and movements, commercialisation, scaling up, reproduction, and policy. Such analysis must move between social and market economy settings. In addition, a policy analysis of institutions currently supporting grassroots innovations will aid our understanding of the ways in which innovation policy can be incorporated.

Sowing seeds of change

Diversity is a key characteristic of resilience and ability to adapt to change. The challenges facing us across the globe demand action both to mitigate, and adapt to environmental, social and economic change. Each of the thematic areas discussed in this book is presently experiencing a crisis: rising oil prices and fears for fuel security, food price rises and shortages, and the threat of economic recession following the 'credit crunch' are signs of tensions in regime systems of provision, prompting greater creativity and innovation in green niches. As mainstream infrastructures of provision are put under pressure, there is greater opportunity for niche solutions to be taken on board by the regime in an attempt to re-stabilise and recover from crisis. Arguably, a diverse range of systems of provision, emerging from innovation in socio-technical systems, and extending beyond the confines of current mainstream institutions and into increasingly self-reliant and empowered communities, will prove the best defence against external shocks. There is a need for diversity and pluralism in social institutions – for a 'better choice of choice' for consumers (Levett *et al.*, 2003). The innovations studied here are specific grassroots responses to the impacts of economic globalisation, as well as environmental risk and social vulnerability. These responses are multi-dimensional, and create space for the expression of different sets of values, objectives and motivations than is possible within the conventional economy. As such, they are valuable experimental, innovative niches, and are the repository of some of the more radical transformative impulses for sustainable consumption driven by ecological citizenship.

It is increasingly evident that the most mundane consumption choices have implications around the globe, and consumers are exhorted to choose responsibly and embrace the political participation opportunities offered with every shopping trip. Ecological citizenship offers a practical, everyday framework for understanding and expressing action which reflects a sense of justice about environmental and social matters through collective efforts to change the institutions which reproduce unsustainable consumption. The policy challenge now is to support those fledgling initiatives seeking to build new institutions for environmental governance, and enable them to grow, thrive and propagate. Supporting alternative development goals and values alongside the familiar market infrastructure is the key to a diverse, robust, adaptable set of social infrastructure and institutions, within which sustainable consumption can be an effective process of change. By combining improvements to the mainstream policy strategy with explicit support for a diversity of alternative approaches which build new social and economic institutions for consumption, governments could harness the energies of ecological citizens working at the grassroots to make significant strides along the road to sustainability.

References

ADAS (2004) *Farmers' Voice 2004 Summary Report: Organic Farming* (Wolverhampton: ADAS Consulting Ltd)

Adger, W. N., Brown, K., Fairbrass, J., Jordan, A., Paavola, J., Rosendo, S. and Seyfang, G. (2003) Governance for Sustainability: Towards a 'Thick' Understanding of Environmental Decision-Making, *Environment and Planning A*, Vol 35(6), pp. 1095–1110

Allen, J. (2006) 'The power to make a difference in an interdependent world', in Smith, J. (ed.) *Interdependence: Mapping the Ethical Terrain of the 21st Century*, p. 8 (Milton Keynes: Open University)

Alliance Data (2007) *Air Miles Reward Program* www.alliancedata.com/oursolutions/airmiles.html, accessed 23/11/07

Amazon Nails (2001) Information guide to straw bale building: for selfbuilders and the construction industry (Todmorden: Amazon Nails)

Amin, A., Cameron, A. and Hudson, R. (2002) *Placing the Social Economy* (London: Routledge)

AMS (2005) *Asda Scoops BBC Award*, News 5/5/05 www.amsltdgroup.com. accessed 21/5/06, copy on file

AMS (2006) *Asda Goes Soft On Local Fruit*, News 25/4/06 www.amsltdgroup.com, accessed 21/5/06, copy on file

Anderson, V. (1991) *Alternative Economic Indicators* (London: Routledge)

Arthur, B. W. (1988) 'Competing technologies: an overview', in Dosi, G., *et al.* (eds) *Technical Change and Economic Theory* (London: Pinter)

Banana Link (2003) *Working Towards Sustainable Trade and Production in Bananas* http://www.bananalink.org.uk

Barnett, C., Cloke, P., Clarke, N. and Malpass, A. (2005) 'Consuming ethics: articulating the subjects and spaces of ethical consumption', in *Antipode*, Vol 37(1), pp. 23–45

Barr, S., Gilg, A. and Shaw, G. (2006) *Promoting Sustainable Lifestyles: A Social Marketing Approach*. Final Summary Report to DEFRA (Exeter: University of Exeter)

Bauman, Z. (2005) *Work, Consumerism and the New Poor* [2nd edition] (Maidenhead: Open University)

Bekin, C., Carrigan, M. and Szmigin, I. (2005) 'Defying marketing sovereignty: voluntary simplicity at new consumption communities', in *Qualitative Market Research: An International Journal*, Vol 8(4), pp. 413–429

Bell, D. (2005) 'Liberal environmental citizenship', in Dobson, A. and Valencia, A. (eds) *Citizenship, Economy and Environment*, pp. 23–38 (London: Routledge)

Berkhout, F. (2002) Technological regimes, path dependency and the environment. *Global Environmental Change*, 12(1): 1–4

Berkhout, F., Smith, A. and A. Stirling (2004) 'Sociotechnical regimes and transition contexts', in Elzen, B., Geels, F. W. and Green, K. (eds) *System*

Innovation and the Transition to Sustainability: Theory, Evidence and Policy (Camberley: Edward Elgar)

Beveridge, R. and Guy, S. (2005) 'The rise of the eco-preneur and the messy world of environmental innovation', *Local Environment*, Vol 10(6), pp. 665–676

Bibbings, J. (2004) *A Sustainability Reward Card for Wales* (Cardiff: Welsh Consumer Council)

Blair, T. (2000) Speech given to Active Community Convention and Awards, 2 March 2000, available at http://www.number10.gov.uk/Page1515, accessed 20/8/08

Blake, J. (1999) 'Overcoming the "value-action gap" in environmental policy: tensions between nation al policy and local experience', *Local Environment*, 4, 257–278

Blunkett, D. (2003) *Active Citizens, Strong Communities: Progressing Civil Renewal* (London: Home Office)

Body Shop, The (1996) *Measuring Up: A Summary of the Body Shop Values Report 1995* (Littlehampton, UK: Ethical Audit, Body Shop International)

Bordieu, P. (1984) *Distinction: A Social Critique of the Judgement of Taste* (London: Routledge)

Born, B. and Purcell, M. (2006) 'Avoiding the local trap: scale and food systems in planning research', *Journal of Planning Education and Research*, 26: 195–207

Boulding, K. (1966) 'The economics of the coming spaceship Earth', in Jarrett, H. (ed.) *Environmental Quality in a Growing Economy* (Baltimore, MD: John Hopkins Press for Resources for the Future)

Boyle, D. (1993) *What Is New Economics?* (London: New Economics Foundation)

Boyle, D. (1996) 'The transatlantic money revolution', in *New Economics Magazine*, No 40 (Winter 1996), pp. 4–7

Boyle, D. (2003) *Beyond Yes And No: A Multi-currency Alternative to EMU* (London: New Economics Foundation)

Boyle, D. (2004) *Co-Production Discussion Paper 1: Work* (London: New Economics Foundation)

Boyle, D. (2005) 'Sustainability and Social Assets', paper presented at the *Grassroots Innovations for Sustainable Development* conference, UCL London, 10th June, 2005, http://www.uea.ac.uk/env/cserge/events/2005/grassroots/index.htm

Boyle, D. (ed.) (2002) *The Money Changers: Currency Reform from Aristotle to E-Cash* (London: Earthscan)

Boyle, G. and P. Harper (1976) *Radical Technology* (London: Wildwood House)

Briceno, T. and Stagl, S. (2006) 'The role of social processes for sustainable consumption', *Journal of Cleaner Production*, 14(17): 1541–1551

Burgess, J., Bedford, T., Hobson, K., Davies, G. and Harrison, C. (2003) '(Un)sustainable consumption', in F. Berkhout, M. Leach and I. Scoones (eds) *Negotiating Environmental Change: New Perspectives from Social Science*, pp. 261–291 (Cheltenham: Edward Elgar)

Burns, S. (2004) *Exploring Co-Production: An Overview of Past, Present and Future* (London: New Economics Foundation)

Burns, S. and Smith, K. (2004) *Co-Production Works! The Win-Win of Involving Local People in Public Services* (London: New Economics Foundation)

Burns, S., Clark, S. and Boyle, D. (2005) *The Downside To Full Employment* (London: New Economics Foundation)

Byrne, D. (2005) *Social Exclusion* [2nd edition] (Maidenhead: Open University Press)

Cahn, E. (2000) *No More Throwaway People: The Co-production Imperative* (Washington: Essential Books)

Cahn, E. and Rowe, J. (1998) *Time Dollars: The New Currency that Enables Americans to Turn their Hidden Resource – Time – into Personal Security and Community Renewal* (Family Resource Coalition of America Chicago) [second edition, first published 1992]

Canelo Project (n.d.) *The Canelo Project: Connecting People, Culture and Nature* (Elgin: Canelo Project)

Carbon Trust (2006) *The Carbon Emissions Generated In All That We Consume* (London: Carbon Trust)

Chanan, G. (2004) *Community Sector Anatomy* (London: Community Development Foundation)

Childs, C. and Whiting, S. (1998) *Eco-Labelling and the Green Consumer*, Sustainable Business Publications Working Paper (Bradford: University of Bradford)

Church, C. (2005) 'Sustainability: The importance of grassroots initiatives', paper presented at the *Grassroots Innovations for Sustainable Development* conference, UCL London, 10th June, 2005, http://www.uea.ac.uk/env/cserge/events/2005/grassroots/index.htm

Church, C. and Elster, J. (2002) *The Quiet Revolution* (Birmingham: Shell Better Britain)

Clark, N. (1985) *The Political Economy of Science and Technology* (Oxford: Basil Blackwell)

Cohen, M. (2001) 'The emergent environmental policy discourse on sustainable consumption', in M. Cohen and J. Murphy (eds) *Exploring Sustainable Consumption: Environmental Policy and the Social Sciences*, pp. 21–37 (London: Pergamon)

Commission on the Future of Volunteering (2008) *Manifesto For Change* (London: Commission on the Future of Volunteering)

Conrad, J. (ed.) (1995) *Eco-Villages and Sustainable Communities: Models for 21st Century Living* (Findhorn: Findhorn Press)

Co-operative Bank (2007) *The Ethical Consumerism Report 2007* (Manchester: Co-operative Bank)

Corborn, J. (2005) *Street Science: Community Knowledge and Environmental Health Justice* (Cambridge, MA: MIT Press)

Corporate Watch (n.d.) *What's Wrong With Supermarkets?* www.corporatewatch.org/?lid=217, accessed 24/5/06, copy on file

Costanza, R. (ed.) (1991) *Ecological Economics: The Science and Management of Sustainability* (New York: Columbia University Press)

Croall, J. (1997) *LETS Act Locally: The Growth of Local Exchange Trading Systems* (London: Calouste Gulbenkian Foundation)

Crompton, T. (2008) *Weathercocks and Signposts: The Environmental Movement at a Crossroads* (London: WWF-UK)

Daly, H. (1992) *Steady State Economics*, 2nd edition (London: Earthscan)

Daly, H. and Cobb, J. (1990) *For The Common Good* (London: Greenprint Press)

Dauncey, G. (1996) *After The Crash: The Emergence of the Rainbow Economy* [2nd edition, first published 1988] (Woodbridge: Greenprint)

Davis-Smith, J. (1998) *The 1997 National Survey of Volunteering* (London: National Centre for Volunteering)

Dawnay, E. and Shah, H. (2005) *Behavioural Economics: Seven Principles for Policy Makers* (London: New Econonomics Foundation)

DCLG (Department for Communities and Local Government) (2007a) *Building A Greener Future: Policy Statement* (London: The Stationery Office)

DCLG (2007b) *Homes For The Future: More Affordable, More Sustainable* (London: The Stationery Office)

DCLG (2008) *The Code for Sustainable Homes: Setting the Standard in Sustainability for New Homes* (London: The Stationery Office)

DEFRA (2002) *The Strategy for Sustainable Farming and Food* (London: DEFRA)

DEFRA (2003a) *Agriculture in the United Kingdom 2003* (London: Stationery Office)

DEFRA (2003b) *Changing Patterns: UK Government Framework for Sustainable Consumption and Production* (London: DEFRA)

DEFRA (2004) *Evidence and Innovation* (London: DEFRA)

DEFRA (2005a) *DEFRA and Social Enterprise: A Position Statement* (London: DEFRA)

DEFRA (2005b) *Delivering Sustainable Development at Community Level*, www.sustainable-development.gov.uk/delivery/global-local/community.htm, accessed 24/10/05

DEFRA (2005c) *Organic Statistics, United Kingdom* (London: DEFRA and National Statistics)

DEFRA (2007a) *A Framework For Pro-Environmental Behaviours* (London: DEFRA)

DEFRA (2007b) *Green Labels and Claims* (London: DEFRA)

DEFRA (2007c) Sustainable Consumption and Production: Our Actions http://www.defra.gov.uk/environment/business/scp/actions/index.htm, accessed 2/08/07, copy on file.

DEFRA (2007d) *Sustainable Development Indicators In Your Pocket* (London: DEFRA)

DEFRA (2008) Climate Change Bill http://www.defra.gov.uk/environment/climatechange/uk/legislation/index.htm, accessed 1/05/08

DeMeulenaere, S. (2004) *Local Exchange Systems in Asia, Africa and Latin America* http://www.appropriate-economics.org, accessed 4/10/04

DETR (Department of the Environment, Transport and the Regions) (1999) *A Better Quality of Life: A Strategy for Sustainable Development for the United Kingdom* (London: DETR)

DETR (2000) *Local Quality of Life Counts: A Handbook for a Menu of Local Indicators of Sustainable Development* (London: DETR)

Devine-Wright, P. (2006) 'Citizenship, responsibility and the governance of sustainable energy systems', in Murphy, J. (ed.) *Framing the Present, Shaping the Future: Contemporary Governance of Sustainable Technologies* (London: Earthscan)

DfT (2007) *Act On CO2* www.dft.gov.uk/ActOnCO2/

DH (Department of Health) (2005) *Independence, Well-Being and Choice: Our Vision for the Future of Social Care for Adults in England*, Green Paper (London: DoH)

Doane, D. (2000) *Corporate Spin: The Troubled Teenage Years of Social Reporting* (London: New Economics Foundation)

Dobson, A. (2003) *Citizenship and the Environment* (Oxford: University Press)

Dobson, A. and Valencia, A. (eds) (2005) *Citizenship, Economy and Environment* (London: Routledge)

Dobson, R. (1993) *Bringing the Economy Home from the Market* (Montreal and New York: Black Rose Books)

Dodd, N. (1994) *The Sociology of Money: Economics, Reason and Contemporary Society* (Cambridge: Polity)

DoE (Department of the Environment) (1993) *Wake Up To What You Can Do For The Environment* (London: DoE)

Dosi, G. (1982) 'Technological paradigms and technological trajectories', *Research Policy*, 11, 147–162

Dosi, G., Freeman, C., Nelson, R., Silverberg, G. and L. Soete (eds) (1988) *Technical Change and Economic Theory* (London: Pinter)

Douglas, M. and Isherwood, B. (1979) *The World of Goods: Towards an Anthropology of Consumption* [second edition, 1996] (London: Routledge)

Douthwaite, R. (1992) *The Growth Illusion* (Bideford, UK: Green Books)

Douthwaite, R. (1996) *Short Circuit: Strengthening Local Economies for Security in an Unstable World* (Totnes, UK: Green Books)

Downing, P. and Ballantyne, P. (2007) *Tipping Point or Turning Point? Social Marketing and Climate Change* (London: Ipsos MORI)

Dresher, E. (1997) 'Valuing Unpaid Work', keynote address, UNPAC's Counting Women's Work Symposium http://www.unpac.ca/economy/valuingunpaid-work.html, accessed 1/08/08

Druckman, A., Bradley, P., Papathanasopoulou, E. and Jackson, T. (2007) 'Measuring progress towards carbon reduction in the UK', *Ecological Economics* (in press) doi: 10.1016/j.ecolecon.2007.10.020

DTI (Department for Trade and Industry) (2002) *Social Enterprise: A Strategy for Success* (London: DTI)

DTI (2003a) *Innovation Report: Competing in the Global Economy, the Innovation Challenge* (London: DTI)

DTI (2003b) *Our Energy Future: Creating A Low-Carbon Economy*, Energy White Paper (Norwich: Stationery Office)

DTI (2005) *Innovation* www.innovation.gov.uk, accessed 24/10/05.

DuPuis, M. and Goodman, D. (2005) 'Should we go "home" to eat?: toward a reflexive politics of localism', *Journal of Rural Studies*, 21(3), 359–371

EAFL (East Anglia Food Link) (2004) About East Anglia Food Link http://www.eafl.org.uk/About EAFL.htm

EAFL (2005) *Local Education Authorities collaborating on local food* http://www.eafl.org.uk/default.asp?topic=SpiceSeven, accessed 10/5/06

Ekins, P. (1992) *A New World Order: Grassroots Movements For Global Change* (London: Routledge)

Ekins, P. (ed.) (1986) *The Living Economy: A New Economics in the Making* (London: Routledge)

Ekins, P. and Max-Neef, M. (eds) (1993) *Real-Life Economics: Understanding Wealth Creation* (London: Routledge)

Ekins, P., Hillman, M. and Hutchinson, R. (1992) *Wealth Beyond Measure: An Atlas of New Economics* (London: Gaia Books)

Elgin, D. (1981) *Voluntary Simplicity: Towards a Way of Life that is Outwardly Simple, Inwardly Rich* (New York: William Morrow)

Energy Saving Trust (2003) The Hockerton Housing Project Energy Efficiency Best Practice in Housing, *New Practice Profile*, 119 (London: Energy Saving Trust)

Energywatch (2008) Cheapest energy bills not the same as affordable bills, says energywatch [press release, 20 March 2008], http://www.energywatch.org.uk

Environment Agency (2008) Water Neutrality http://www.environment-agency. gov.uk/subjects/waterres/287169/1917628/?lang=_e, accessed 31/03/08

Environmental Audit Committee (2003) Learning the Sustainability Lesson, 10th Report of Session 2002–2003 (London: Stationery Office)

Eostre Organics (2004) *The Eostre Organics Charter* http://www.eostreorganics. co.uk/charter.htm, accessed 30/3/04, copy on file

ETI (Ethical Trade Initiative) (2003a) *ETI Statement on Littlewoods and Ethical Trading Initiative Membership*, press release 24 January 2003, http://www.ethicaltrade.org/pub/publications/2003/01-stmt-ltlwds/index.shtml

ETI (2003b) *What Is ETI?* http://www.ethicaltrade.org/pub/about/eti/main/index.shtml, accessed 5/2/03

Etzioni, A. (1998) 'Voluntary simplicity: characterisation, select psychological implications and societal consequences', in *Journal of Economic Psychology*, Vol 19(5), pp. 619–643

European Commission (2004) Sustainable consumption and production in the European Union (Luxembourg: Office for Official Publications of the European Communities)

European Council (2006) 'Review of the EU Sustainable Development Strategy (SDS) – Renewed Strategy', 10117/06 Annex (Luxembourg: Council of the European Union)

Fairlie, S. (1996) *Low-Impact Development: Planning and People in a Sustainable Countryside* (Charlbury: Jon Carpenter Publishing)

FARMA (2006) *Farmers' Markets in The UK: Nine Years and Counting*, Sector Briefing (Southampton: National Farmers Retail and Markets Association)

Festinger, L. (1957) *A Theory of Cognitive Dissonance* (Stanford: University of California Press)

Filkin, G., Stoker, G., Wilkinson, G. and Williams, J. (2000) *Towards a New Localism* (London: New Local Government Network)

Fine, B. (2002) *The World of Consumption: The Material and Cultural Revisited* [second edition] (London: Routledge)

Fine, B. and Leopold, E. (1993) *The World of Consumption* (London: Routledge)

Fleming, D. (2005) *Energy and the Common Purpose: Descending the Energy Staircase with Tradable Energy Quotas (TEQs)*. (London: The Lean Economy Connection)

Frey, B. and Jegen, R. (2001) 'Motivation Crowding Theory: A survey of empirical evidence', in *Journal of Economic Surveys*, Vol 15(5), pp. 589–611

Fussler, C. and James, P. (1996) *Driving Eco-Innovation: A Breakthrough Discipline for Innovation and Sustainability* (London: Pitman)

Geels, F. W. (2004) 'From sectoral systems of innovation to sociotechnical systems. Insights about dynamics and change from sociology and institutional theory', *Research Policy*, 33, 897–920

Georg, S. (1999) 'The social shaping of household consumption', in *Ecological Economics*, Vol 28, pp. 455–466

Gesell, S. (1958) *The Natural Economic Order* (London: Peter Owen)

Ghazi, P. and Jones, J. (1997) *Getting A Life: The Downshifter's Guide to Happier, Simpler Living* (London: Hodder and Stoughton)

Gibson-Graham, J. K. (1996) *The End of Capitalism (as we knew it)* (Cambridge MA: Blackwell)

Giddens, A. (1984) *The Constitution of Society: An Outline Of The Theory Of Structuration* (Cambridge: Polity Press)

Giddens, A. (1998) *The Third Way: The Renewal of Social Democracy* (Cambridge: Polity Press)

Gladwell, M. (2000) *The Tipping Point: How Little Things Can Make a Big Difference* (London: Abacus)

Gonella, C. and Pilling, A. (1998) *Making Values Count: Contemporary Experience in Social and Ethical Accounting, Auditing and Reporting*, Research Report 57 (London: Certified Accountants Educational Trust)

Goodall, C. (2007) *How To Live A Low-Carbon Life: The Individual's Guide to Stopping Climate Change* (London: Earthscan)

Gorz, A. (1999) *Reclaiming Work* (Cambridge: Polity Press)

Greco, T. (1994) *New Money For Healthy Communities*, T. Greco, PO Box 42663, Tucson, Arizona 85733

Green Consumer Guide.com (2007) *Petrol, Diesel and alternative-fuelled vehicles* http://www.greenconsumerguide.com/transport_main.php, accessed 19/02/08

Guthman J. (2003) 'Fast Food / Organic Food: reflexive tastes and the making of "yuppie chow"', *Social and Cultural Geography*, 4(1), 45–58

Guy, S. (1997) *Alternative Developments: The Social Construction of Green Buildings*, Cutting Edge 97 (London: RICS)

Hacker, J., Belcher, S. and Connell, R. (2005) *Beating the Heat: Keeping UK Buildings Cool in a Warming Climate*, UKCIP Briefing Report (Oxford: UKCIP)

Hajer, M. (1995) *The Politics of Environmental Discourse: Ecological Modernization and the Policy Process* (Oxford: Clarendon Press)

Hart, K. (n.d.) *A Sampler of Alternative Homes: Approaching Sustainable Architecture* VHS (Crestone, Co.: Hartworks) available from www.greenhomebuilding.com

Hassanein, N. (2003) 'Practicing food democracy: a pragmatic politics of transformation', *Journal of Rural Studies*, 19, 77–86

Heap, B. and Kent, J. (eds) (2000) *Towards Sustainable Consumption: A European Perspective* (London: Royal Society)

Henderson, H. (1995) *Paradigms in Progress: Life Beyond Economics* [originally published 1991] (San Francisco: Berrett-Koehler Publishers)

Henderson, H. (1996) 'Changing paradigms and indicators: implementing equitable, sustainable and participatory development', in Griesgraber, J. and Gunter, B. (1996) *Development* (London: Pluto Press)

Henderson, H. and Ikeda, D. (2004) *Planetary Citizenship: Your Values, Beliefs and Actions Can Shape a Sustainable World!* (Santa Monica, CA: Middleway Press)

Hewitt, M. and Telfer, K. (2007) *Earthships: Building a Zero Carbon Future for Homes* (Bracknell: IHE BRE Press)

Hillman, M. (2004) *How We Can Save the Planet* (London: Penguin)

Hines, J. (2005) 'Grassroots initiatives in housing', paper presented at the *Grassroots Innovations for Sustainable Development Conference*, 10th June, 2005 (London: UCL), http://www.uea.ac.uk/env/cserge/events/2005/grassroots/index.htm

Hinrichs, C. C. (2003) 'The practice and politics of food system localization', *Journal of Rural Studies*, Vol 19, pp. 33–45

Hirsch, F. (1977) *Social Limits to Growth* (London: Routledge)

HM Government (2005) *Securing The Future: Delivering UK Sustainable Development Strategy* (Norwich: The Stationery Office)

HM Treasury (2002a) *Exploring the Role of the Third Sector in Public Service Delivery and Reform: A Discussion Document* (London: HM Treasury)

HM Treasury (2002b) *The Role of the Voluntary and Community Sector in Service Delivery: A Cross-cutting Review* (London: HM Treasury)

Hobson, K. (2002) 'Competing discourses of sustainable consumption: Does the "rationalisation of lifestyles" make sense?', in *Environmental Politics*, Vol 11(2), pp. 95–120

Hobson, K. (2001) 'Sustainable lifestyles: rethinking barriers and behaviour change', in M. Cohen and J. Murphy (eds) *Exploring Sustainable Consumption: Environmental Policy and the Social Sciences*, pp. 191–209 (London: Pergamon)

Holdsworth, M. (2003) *Green Choice: What Choice?* (London: National Consumer Council)

Holdsworth, M. and Boyle, D. (2004) *Carrots Not Sticks: The Possibilities of a Sustainability Reward Card for the UK* (London: New Economics Foundation and National Consumer Council)

Holliday, C. and Pepper, J. (2001) *Sustainability Through the Market: Seven Keys to Success* (Geneva: WBCSD)

Holloway, L. and Kneafsey, M., (2000) 'Reading the space of the farmer's market: A case study from the United Kingdom', *Sociologica Ruralis*, 40, 285–299

Home Office (1998) Compact on Relations Between Government and the Voluntary and Community Sector (London: Home Office) http://www.dwp.gov.uk/housing benefit/ manuals/hbgm/parts/ptc_03b.asp

Hoogma, R., Kemp, R., Schot, J. and B. Truffer (2002) *Experimenting for Sustainable Transport: The Approach of Strategic Niche Management* (London: Spon Press)

Hoskyns, C. and Rai, S. (2007) 'Recasting the global political economy: counting women's unpaid work', *New Political Economy*, Vol 12(3), pp. 297–317

Hughes, T. P. (1983) *Networks of Power: Electrification in Western Society, 1880–1930* (Baltimore: Johns Hopkins University Press)

198   *References*

Hulme, M., Jenkins, G., Lu, X., Turnpenny, J., Mitchell, T., Jones, R., Lowe, J., Murphy, J., Hassell, D., Boorman, P., McDonald, R. and Hill, S. (2002) *Climate Change Scenarios for the United Kingdom: The UKCIP02 Scientific Report* (Norwich: Tyndall Centre for Climate Change Research, UEA)

Hutchinson, F., Mellor, M. and Olsen, W. (2002) *The Politics of Money: Towards Sustainability and Economic Democracy* (London: Pluto)

IGD (2003) 'Local food comes from our county, say consumers', Press release 1/5/03 available at www.igd.com, accessed 6/5/04, copy on file

IPCC (2007) Climate Change 2007: Synthesis Report. Contribution of Working Groups I, II and III to the Fourth Assessment Report of the Intergovernmental Panel on Climate Change [Core Writing Team, Pachauri, R. K. and Reisinger, A. (eds)]. (Geneva, Switzerland: IPCC)

Irwin, A., Georg, S. and P. Vergragt (1994) 'The social management of environmental change', *Futures*, 26(3), pp. 323–334

Jackson, T. (2004a) *Chasing Progress: Beyond Measuring Economic Growth* (London: New Economics Foundation)

Jackson, T. (2004b) *Models of Mammon: A Cross-Disciplinary Survey in Pursuit of the 'Sustainable Consumer'*, ESRC Sustainable Technologies Programme Working Paper No 2004/1 (Guildford: Centre for Environmental Strategy)

Jackson, T. (2005) *Motivating Sustainable Behaviour: A Review of Evidence on Consumer Behaviour and Behaviour Change*, Report to the Sustainable Development Research Network (London: Policy Studies Institute)

Jackson, T. (2007a) 'Readings in sustainable consumption', in T. Jackson (ed.) *The Earthscan Reader in Sustainable Consumption*, pp. 1–23 (London: Earthscan)

Jackson, T. (2007b) 'Consuming Paradise? Towards a social and cultural psychology of sustainable consumption', in T. Jackson (ed.) *The Earthscan Reader in Sustainable Consumption*, pp. 367–395 (London: Earthscan)

Jackson, T. and Marks, N. (1994) *Measuring Sustainable Economic Welfare – A Pilot Index: 1950–1990* (London and Stockholm: New Economics Foundation and Stockholm Environment Institute)

Jackson, T. and Marks, N. (1999) 'Consumption, sustainable welfare and human needs – with reference to UK expenditure patterns between 1954 and 1994', *Ecological Economics*, 18, pp. 421–441

Jackson, T. and Michaelis, L. (2003) *Policies for Sustainable Consumption* (Oxford: Sustainable Development Commission)

Jackson, T. and Michaelis, L. (2003) *Policies For Sustainable Consumption* (Oxford: Sustainable Development Commission)

Jacobs, J. (1984) *Cities and the Wealth of Nations: Principles of Economic Life* (London: Random House)

Jacobsson, S. and Johnson, A. (2000) 'The diffusion of renewable energy technology: an analytical framework and key issues for research', *Energy Policy*, 28, 625–640

Jaeger, C., Dürrenberger, G., Kastenholz, H. and Truffer, B. (1993) 'Determinants of environmental action with regard to climate change', in *Climate Change*, 23, 193–211

Jasanoff, S. and Martello, M. (2004) *Earthly Politics: Local and Global in Environmental Governance* (Cambridge, MA: MIT Press)

Jones, A. (2001) *Eating Oil: Food Supply in a Changing Climate* (Newbury: Sustain, London and Elm Farm Research Centre)

Kahneman, D., Knetsch, J. and Thaler, R. (1991) 'Anomalies: the endowment effect, loss aversion and status quo bias', *Journal of Economic Perspectives*, Vol 5(1), pp. 193–206

Kemp, R., Schot, J. and Hoogma, R. (1998) 'Regime shifts to sustainability through processes of niche formation: the approach of strategic niche management', *Technology Analysis and Strategic Management*, 10(2), 175–195

Kemp, S. and Cowie, P. (2004) *The Earthship Toolkit: Your Guide to Building a Zero Waste, Zero Energy Future* (Kinghorn: Sustainable Communities Initiatives)

Kendall, J. and Almond, S. (1999) 'United Kingdom', in Salamon, L. M., Anheier, H. K., List, R., Toepler, S., Sokolowski, S. W., and Associates (eds) *Global Civil Society: Dimensions of the Nonprofit Sector, Volume One* (Baltimore: John Hopkins Centre for Civil Society Studies)

Keynes, J. M. (1973) *The General Theory of Employment, Interest and Money* [first published 1936] (London: Macmillan)

Kolmuss, A. and Agyeman, J. (2002) 'Mind the gap: why do people act environmentally and what are the barriers to pro-environmental behavior?', *Environmental Education Research*, Vol 8, No 3, pp. 239–260

Kotler, P. and Zaltman, G. (1971) 'Social marketing: an approach to planned social change', *Journal of Marketing*, Vol. 35(3), pp. 3–12

Lang, P. (1994) *LETS Work: Rebuilding the Local Economy* (Bristol, UK: Grover Books)

Law, B. (2005) *The Woodland House* (East Meon, Hampshire: Permanent Publications)

Lee, R., Leyshon, A., Aldridge, T., Tooke, J., Williams, C. C. and Thrift, N. (2004) 'Making geographies and histories? Constructing local circuits of value', *Environment and Planning D: Society and Space*, Vol 22, No 4, pp. 595–617

Levett, R., with Christie, I., Jacobs, M. and Therivel, R. (2003) *A Better Choice of Choice: Quality of Life, Consumption and Economic Growth* (London: Fabian Society)

Leyshon, A., Lee, R. and Williams, C. (eds) (2003) *Alternative Economic Spaces* (London: Sage)

Lietaer, B. (2001) *The Future of Money: Creating New Wealth, Work and a Wiser World* (London: Century)

Lindsay, I. (2001) 'The voluntary sector', in Crick, B. (ed.) *Citizens: Towards a Citizenship Culture* (Oxford: Blackwell)

Linton, M. and Soutar, A. (1996) *Money and the Sustainable Economy* http://www.gmlets.u-net.com/explore/sustain.html, accessed 23/11/07

Lipsey, R. and Harbury, C. (1992) *First Principles of Economics*, 2nd edition (Oxford: Oxford University Press)

Lodziak, C. (2002) *The Myth of Consumerism* (London: Pluto)

Lovell, H. (2004) 'Framing sustainable housing as a solution to climate change', *Journal of Environmental Policy and Planning*, Vol 6(1), pp. 35–55

Lovell, H. (2005) 'Supply and demand for low-energy housing in the UK: insights from a science and technology studies approach', *Housing Studies*, Vol 20(5), pp. 815–829

Lutz, M. (1999) *Economics for the Common Good: Two Centuries of Social Economic Thought in the Humanistic Tradition* (London: Routledge)

Lutz, M. and Lux, K. (1988) *Humanistic Economics: The New Challenge* (New York: Bootstrap Press)

Lynas, M. (2007) 'Can shopping save the planet?', *The Guardian G2* 17/9/07 pp. 4–7

MacGillivray, A., Weston, C. and Unsworth, C. (1998) *Communities Count! A Step-by-Step Guide to Community Sustainability Indicators* (London: New Economics Foundation)

Maniates, M. (2002) 'Individualization: plant a tree, buy a bike, save the world?', in T. Princen, M. Maniates and K. Konca (eds) *Confronting Consumption*, pp. 43–66 (London: MIT Press)

Manno, J. (2002) 'Commoditization: consumption efficiency and an economy of care and connection', in T. Princen, M. Maniates and K. Konca (eds) *Confronting Consumption*, pp. 67–99 (London: MIT Press)

Marks, N., Simms, A., Thompson, S. and Abdallah, S. (2006) *The Happy Planet Index: An Index of Human Wellbeing and Environmental Impact* (London: New Economics Foundation)

Maslow, H. (1954) *Motivation and Personality* (New York: Harper and Row)

Max-Neef, M. (1992) 'Development and human needs', in Ekins, P. and Max-Neef, M. (eds) *Real-Life Economics: Understanding Wealth Creation*, pp. 197–213 (London: Routledge)

Maynard, R. and Green, M. (2006) *Organic Works* (Bristol: Soil Association)

McDonald, S., Oates, C., Young, W. and H. Kumju (2006) 'Toward sustainable consumption: researching voluntary simplifiers', *Psychology and Marketing*, Vol 23(6), pp. 515–534

McKenzie-Mohr, D. and Smith, W. (1999) *Fostering Sustainable Behaviour: An Introduction to Community-based Social Marketing* (Philadelphia PA: New Society Publishers)

Meadows, D. H., Meadows, D. L., Randers, J. and Behrens, W. W. (1972) *The Limits To Growth* (New York: Universe Books)

Meeker-Lowry, S. (1995) *Invested in the Common Good* (Philadelphia PA: New Society Publishers)

Meltzer, G. (2005) *Sustainable Community: Learning From the Co-housing Model* (Trafford: Crewe)

Mesure, S. (2005) 'Asda plans to double share of local produce', *The London Independent*, Sept 28, 2005

Miller, D. (ed.) (1995) *Acknowledging Consumption: A Review of New Studies* (London: Routledge)

Monbiot, G. (2007) 'Ethical Shopping Is Just Another Way Of Showing How Rich You Are', *Guardian* July 24, 2007, p. 27

Morgan, K. and Morley, A. (2002) Relocalising the Food Chain: The role of creative public procurement (Cardiff: The Regeneration Institute)

Moulaert, F. and Ailenei, O. (2005) 'Social economy, third sector and solidarity relations: a conceptual synthesis from history to present', *Urban Studies*, Vol 42(11), pp. 2037–2053

Mulgan, G., Ali, R., Halkett, R. and Sanders, B. (2007) *In and Out of Sync: The Challenge of Growing Social Innovations* (London: NESTA)

Mulgan, G., Wilkie, N., Tucker, S., Ali, R., David, F. and Liptrot, T. (2006) *Social Silicon Valleys: A Manifesto for Social Innovation* (London: Young Foundation)

Murdoch, J., Marsden, T. and Banks, J. (2000) 'Quality, nature and embeddedness', *Economic Geography*, Vol 76(2), pp. 107–125

Murphy, J. (2000) 'Ecological modernisation', *Geoforum*, Vol 31(1), pp. 1–8

Nash, V. and Paxton, W. (2002) *Any Volunteers for the Good Society? Volunteering and Civic Renewal* (London: IPPR)

NEF (2003) *New Survey Launched At Localism Conference...* press release, 16 May 2003, available at www.neweconomics.org

Nelson, R. R. and Winter, S. G. (1982) *An Evolutionary Theory of Economic Change* (Cambridge, Mass.: Bellknap Press)

New Economics Foundation (1995) *Social Statement 1994–95* (London: New Economics Foundation)

Norberg-Hodge, H., Merrifield, T., and Gorelick, S. (2000) *Bringing The Food Economy Home: The Social, Ecological and Economic Benefits of Local Food* (Dartington: ISEC)

North, P. (2006) *Alternative Currency Movements as a Challenge to Globalisation? A Case Study of Manchester's Local Currency Networks* (Aldershot: Ashgate)

O'Riordan, T. (2001) *Globalism, Localism and Identity: Fresh Perspectives on the Sustainability* (London: Earthscan)

OECD (2002a) *Policies to Promote Sustainable Consumption: An Overview*, ENV/EPOC/WPNEP(2001)18/FINAL (Paris: OECD)

OECD (2002b) *Towards Sustainable Consumption: An Economic Conceptual Framework*, ENV/EPOC/WPNEP(2001)12/FINAL (Paris: OECD)

ONS (Office of National Statistics) (2003) *Regional Trends 38* (London: The Stationery Office)

Padbury, G. (2006) *Retail and Foodservice Opportunities for Local Food* (Watford: IGD)

Pearce, D. and Turner, R. K. (1990) *Economics of Natural Resources and the Environment* (Hemel Hempstead, UK: Harvester Wheatsheaf)

Pearson, D. (1989) *The Natural House Book: Creating a Healthy, Harmonious and Ecologically Sound Home* (London: Conran Octopus)

Pearson, R. (2003) Argentina's barter network: New currency for new times?, *Bulletin of Latin American Research*, Vol 22(2), pp. 214–230

Pearson, R. and Seyfang, G. (2001) 'New dawn or false hope? Codes of conduct and social policy in a globalising world', *Global Social Policy*, Vol 1, No 1, pp. 49–79

Pepper, D. (1996) *Modern Environmentalism: An Introduction* (London: Routledge)

PIU (2002) *Social Capital: A Discussion Paper* (London: PIU)

Porritt, J. (2003) *Redefining Prosperity: Resource Productivity, Economic Growth and Sustainable Development* (London: Sustainable Development Commission)

Pretty, J. (2001) *Some Benefits and Drawbacks of Local Food Systems*, briefing note for TVU/Sustain AgriFood Network, November 2, 2001

Pretty, J. (2002) *Agri-Culture: Reconnecting People, Land and Nature* (London: Earthscan)

Pretty, J., Ball, A. S., Lang, T. and Morison, J. I. L. (2005) 'Farm costs and food miles: An assessment of the full cost of the UK weekly food basket', *Food Policy*, Vol 20, pp. 1–19

Princen, T. (2002a) 'Consumption and its externalities: where economy meets ecology', in T. Princen, M. Maniates and K. Konca (eds) *Confronting Consumption*, pp. 23–42 (London: MIT Press)

Princen, T. (2002b) 'Distancing: consumption and the severing of feedback', in T. Princen, M. Maniates and K. Konca (eds) *Confronting Consumption*, pp. 103–131 (London: MIT Press)

Reed, M. (2001) 'Fight The Future! How the contemporary campaigns of the UK organic movement have arisen from their composting past', *Sociologica Ruralis*, Vol 41(1), pp. 131–145

Renting, H., Marsden, T. and Banks, J. (2003) 'Understanding alternative food networks: exploring the role of short food supply chins in rural development', *Environment and Planning A*, Vol 35, pp. 393–411

Reynolds, M. (1990) *Earthship Volume I* (Taos: Solar Survival Press)

Reynolds, M. (2000) *Comfort in any Climate* (Taos: Solar Survival Press)

Reynolds, M. (2004a) 'Earthships' presentations given at *Earthship Seminar*, September 17–19, 2004, Taos, New Mexico

Reynolds, M. (2004b) 'Introduction to Earthships', presentation given at *International Earthship Summit* October 29–31 2004 Brighton University, Brighton

Ricketts Hein, J., Ilbery, B. and Kneafsey, M. (2006) 'Distribution of local food activity in England and Wales: An index of food relocalisation', *Regional Studies*, Vol 40(3), pp. 289–301

Rip, A. and Kemp, R. (1998) 'Technological change', in Rayner, S. and E. L. Malone (eds) *Human Choice and Climate Change, Volume 2* (Columbus: Battelle Press)

Roberts, S. (2005) 'Grassroots initiatives in energy', paper presented at the *Grassroots Innovations for Sustainable Development* conference, UCL London, 10th June, 2005, http://www.uea.ac.uk/env/cserge/events/2005/grassroots/index.htm

Robertson, J. (1985) *Future Work: Jobs, Self-employment and Leisure After the Industrial Age* (Aldershot, UK: Gower)

Robertson, J. (1999) *The New Economics of Sustainable Development: A Briefing for Policymakers* (London: Kogan Page)

Rogers, B. and Robinson, E. (2004) *The Benefits of Community Engagement: A Review of the Evidence* (London: Home Office)

Røpke, I. (1999) 'The dynamics of willingness to consume', *Ecological Economics*, Vol 28, pp. 399–420

Røpke, I. and Reisch, L. (2004) 'The place of consumption in Ecological Economics', in Reisch, L. and Røpke, I. (eds) *The Ecological Economics of Consumption*, pp. 1–16 (Cheltenham: Edward Elgar)

Russell, S. and Williams, R. (2002) 'Social shaping of technology: frameworks, findings and implications for policy with glossary of social shaping concepts', in Sørensen, K.H. and R. Williams (eds) *Shaping Technology, Guiding Policy: Concepts, Spaces and Tools* (Cheltenham: Edward Elgar)

Sagoff, M. (1988) 'Some problems with environmental economics', in *Environmental Ethics*, Vol 10, No 1, pp. 55–74

Sale, K. (1980) *Human Scale* (London: Secker & Warburg)

Saltmarsh, N. (2004) *Mapping the Food Supply Chain in the Broads and Rivers Area* (Watton: East Anglia Food Link)

Sanne, C. (2002) 'Willing consumers – or locked-in? Policies for a sustainable consumption', *Ecological Economics*, Vol 42, pp. 273–287

Schellnhuber, H., Cramer, W., Nakicenovic, N., Wigley, T. and Yohe, G. (eds) (2006) *Avoiding Dangerous Climate Change* (Cambridge: Cambridge University Press)

Schlich, E. and Fleissner, U. (2005) 'The ecology of scale: assessment of regional turnover and comparison with global food', *International Journal of Life Cycle Assessment*, Vol 10(3), pp. 168–170

Schor, J. (1998) *The Overspent American: Upscaling, Downshifting and the New Consumer* (New York: Basic Books)

Schot, J. 1998, 'The usefulness of evolutionary models for explaining innovation: the case of the Netherlands in the 19th century', *History and Technology*, 14, 173–200

Schot, J., Hoogma, R. and Elzen, B. (1994) 'Strategies for shifting technological systems: the case of the automobile system', *Futures*, 26(10), 1060–1076

Schumacher, E. F. (1993) *Small Is Beautiful: A Study of Economics as if People Mattered* [first published 1973] (London: Vintage)

Schumpeter, J. (1961) *The Theory of Economic Development* (Oxford: Oxford University Press)

Schwarz, B. (2004) *The Paradox of Choice: Why More is Less* (New York: Harper-Collins)

Schor, J. (1991) *The Overworked American: The Unexpected Decline of Leisure* (New York: Basic Books)

SCR (Sustainable Consumption Roundtable) (2006) *I Will If You Will: Towards Sustainable Consumption* (London: Sustainable Development Commission and National Consumer Council)

SDC (Sustainable Development Commission) (2001) *Unpacking Sustainable Development* http://www.sd-commission.gov.uk/commission/plenary/apr01/unpack/index.htm, accessed 28/10/02, copy on file

SDC (2006) *Stock Take: Delivering Improvements in Existing Housing* (London: SDC)

Seyfang, G. (1999) 'Making it count: valuing and evaluating community economic development', in Haughton, G. (ed.) *Community Economic Development*, pp. 125–138, 244–247 (London: Stationery Office and Regional Studies Association)

Seyfang, G. (2000) 'The Euro, the Pound and the shell in our pockets: rationales for complementary currencies in a global economy', *New Political Economy*, Vol 5, No 2, pp. 227–246

Seyfang, G. (2001a) 'Community currencies: small change for a green economy', *Environment and Planning A*, Vol 33(6), pp. 975–996

Seyfang, G. (2001b) Money that makes a change: community currencies North and South, *Gender and Development*, Vol 9, No 1, pp. 60–69

Seyfang, G. (2001c) Working for the Fenland dollar: An evaluation of Local Exchange Trading Schemes (LETS) as an informal employment strategy to tackle social exclusion, *Work, Employment and Society*, Vol 15(3), pp. 581–593

Seyfang, G. (2002) 'Tackling social exclusion with community currencies: learning from LETS to time banks', *International Journal of Community Currency Research*, Vol 6 <http://www.uea.ac.uk/env/ijccr/>

Seyfang, G. (2003) 'Growing cohesive communities, one favour at a time: social exclusion, active citizenship and time banks', *International Journal of Urban and Regional Research*, Vol 27(3), pp. 699–706

Seyfang, G. (2004a) 'Consuming values and contested cultures: a critical analysis of the UK strategy for sustainable consumption and production', *Review of Social Economy*, Vol 62(3), pp. 323–338

Seyfang, G. (2004b) 'Time banks: rewarding community self-help in the inner city?,' *Community Development Journal*, Vol 39(1), pp. 62–71

Seyfang, G. (2004c) 'Working outside the box: community currencies, time banks and social inclusion', *Journal of Social Policy*, Vol 33(1), pp. 49–71

Seyfang, G. (2005) 'Shopping for sustainability: can sustainable consumption promote ecological citizenship?', *Environmental Politics*, Vol 14(2), pp. 290–306

Seyfang, G. (2006a) 'Ecological citizenship and sustainable consumption: examining local food networks', *Journal of Rural Studies*, Vol 40(7) pp. 781–791

Seyfang, G. (2006b) 'Harnessing the potential of the social economy? Time Banks and UK public policy', *International Journal of Sociology and Social Policy*, Vol 26 (9/10), pp. 430–443

Seyfang, G. (2006c) 'New institutions for sustainable consumption: an evaluation of community currencies', *Regional Studies*, Vol 40(7), pp. 781–791

Seyfang, G. (2007) 'Cultivating carrots and community: local organic food and sustainable consumption', *Environmental Values*, Vol 16, pp. 105–123

Seyfang, G. and Smith, A. (2007) 'Grassroots innovations for sustainable development: towards a new research and policy agenda', *Environmental Politics*, Vol 16(4), pp. 584–603

Seyfang, G. and Smith, K. (2002) *The Time of Our Lives: Using Time Banking for Neighbourhood Renewal and Community Capacity-building* (London: New Economics Foundation)

Seyfang, G. with Eastaugh, D., Lees, M., Alexander, G., Seekings, S. and Fanning, M. (eds) (2005) *The Rising Sun: Celebrating Dance Camp East* (Norwich: Dance Camp East)

Seyfang, G., Lorenzoni, I. and Nye, M. 'Personal carbon trading: a critical examination of proposals for the UK', *Ecological Economics*, paper submitted to Ecological Economics

Shah, H. and Marks, N. (2004) *A Well-being Manifesto for a Flourishing Society* (London: New Economics Foundation)

Shaw, D., Newholm, T. and Dickinson, R. (2006) 'Consumption as voting: an exploration of consumer empowerment', *European Journal of Marketing*, Vol 40(9/10), pp. 1049–1067

Shove, E. (1998) 'Gaps, barriers and conceptual chasms: theories of technology transfer and energy in buildings', *Energy Policy*, Vol 26(15), pp. 1105–1112

Shove, E. (2004) 'Changing human behaviour and lifestyle: a challenge for sustainable consumption?', in Reisch, L. and Røpke, I. (eds) *The Ecological Economics of Consumption*, pp. 111–131 (Cheltenham: Edward Elgar)

Shove, E. (2003) *Comfort, Cleanliness and Convenience: The Social Organization of Normality* (Oxford: Berg Publishers)

Simms, A. (2003) *An Environmental War Economy: The Lessons of Ecological Debt and Global Warming* (London: NEF)

Simms, A., Moran, D. and Chowla, P. (2006) *The UK Interdependence Report: How the World Sustains the Nation's Lifestyles and the Price it Pays* (London: New Economics Foundation)

Smith, A. (2004) 'Alternative technology niches and sustainable development', *Innovation: Management, Policy and Practice*, 6(2), 220–235

Smith, A. (2006) 'Green niches in sustainable development: the case of organic food in the UK', *Environment and Planning C*, Vol 24, pp. 439–458

Smith, A. (2007) 'Translating sustainabilities between green niches and socio-technical regimes', *Technology Analysis & Strategic Management*, Vol 19(4), pp. 427–450

Smith, A., Stirling, A. and F. Berkhout (2005) 'The governance of sustainable socio-technical transitions', *Research Policy*, 34, 1491–1510

Smith, A., Watkiss, P., Tewddle, G., McKinnon, A., Browne, M., Hunt, A., Treleven, C., Nash, C. and Cross, S. (2005) *The Validity of Food Miles as an Indicator of Sustainable Development* (London: DEFRA)

Smith, E. and Marsden, T. (2004) 'Exploring the "limits to growth" in UK organics: beyond the statistical image', *Journal of Rural Studies*, Vol 20, pp. 345–357

Soil Association (2003) *Organic Food And Farming Report 2002* (Association: Soil, Bristol)

Soil Association (2005a) *Organic Food and Farming Report 2004* (Bristol: Soil Association)

Soil Association (2005b) *Organic Market Report* (Bristol: Soil Association)

Southerton, D., Chappells, H. and Van Vliet, V. (2004) *Sustainable Consumption: The Implications of Changing Infrastructures of Provision* (Aldershot: Edward Elgar)

Spaargaren, G. (2003) 'Sustainable consumption: a theoretical and environmental policy perspective', *Society and Natural Resources*, 16, 687–701

Steen, A. and Steen, B. (2001) *The Beauty of Straw Bale Homes* (Totnes: Chelsea Green)

Steen, A., Steen, B., Bainbridge, D. with Eisenberg, D. (1994) *The Straw Bale House* (Totnes: Chelsea Green)

Steffen, A. (2007) Strategic consumption: how to change the world with what you buy, <http://www.worldchanging.com/archives/006373.html> accessed 26/3/07

Stern, N. (2007) *The Economics of Climate Change: The Stern Review* (Cambridge: Cambridge University Press)

Stirling, A. (1998) 'On the economics and analysis of diversity', *SPRU ElectronicWorking Paper Series No. 28* (available at: http://www.sussex.ac.uk/spru/1-6-1-2-1.html)

Stirling, A. (2004) 'Opening up or closing down: analysis, participation and power in the social appraisal of technology', in Leach, M., Scoones, I. and Wynne B. (eds) *Science, Citizenship and Globalisation* (London: Zed)

Stringer, E. (1996) *Action Research: A Handbook for Practitioners* (London: Sage)

Sustainable Development Commission (2004) Healthy Futures #2 (London: SDC)

Sustainable Seattle (1993) *Indicators of Sustainable Community: A Report to Citizens on Long-Term Trends in Our Community* (Seattle, USA: Sustainable Seattle)

Swann, R. (1981) 'The place of a local currency in a world economy: towards an economy of permanence', paper presented at the Schumacher Society conference *Community Survival in the Age of Inflation*, April 25th 1981, available at http://www.smallisbeautiful.org/publications/essays_swann/place_of_local_currency.html, accessed 21/2/08

Tallontire, A. and Blowfield, M. (2000) 'Will the WTO prevent the growth of ethical trade? Implications of potential changes to WTO rules for environmental and social standards in the forest sector', *Journal of International Development*, 12, pp. 571–584

Taylor, J., Madrick, M. and Collin, S. (2005) *Trading Places: The Local Economic Impact of Street Produce and Farmers' Markets* (London: New Economics Foundation)

Tegtmeier, E. and Duffy, M. (2004) 'The external costs of agricultural production in the United States', *International Journal of Agricultural Sustainability*, Vol 2, pp. 55–175

Thompson, C. J. and Arsel, Z. (2004) 'The Starbucks brandscape and consumers' anticorporate experiences of globalization', *Journal of Consumer Research*, Vol 31, pp. 631–642

Thompson, S., Abdallah, S., Marks, N., Simms, A. and Johnson, V. (2007) *The European Happy Planet Index: An Index of Carbon Efficiency and Wellbeing in the EU* (London: New Economics Foundation)

Tibbett, R. (1997) 'Alternative currencies: A challenge to globalisation?', *New Political Economy*, Vol 2, No 1, pp. 127–135

Time Banks UK (2001) Statement of Values (Time Banks UK, Gloucester)

Time Banks UK (2005) Welcome To Time Banks UK http://www.timebanks.co.uk, accessed 11/10/05

Time Banks UK (2006) Time Banking and State Benefits http://www.timebanks.co.uk/timebankingandbenefits.asp, accessed 26/1/06, copy on file.

Time Banks UK (2007) Self-Assessment Questionnaire http://www.timebanks.co.uk/Self_assessment_Questionaire.asp, accessed 23/11/07

Titmuss, R. (1970) *The Gift Relationship* (London: Allen and Unwin)

Traidcraft plc (1994) *Traidcraft plc Social Audit 1993/94* (Gateshead, UK: Traidcraft plc)

UN (2002) Johannesburg Plan of Implementation (New York: United Nations)

UN Division for Sustainable Development (2007) '10-year framework of programmes on sustainable consumption and production patterns: the Marrakech Process' http://www.un.org/esa/sustdev/sdissues/consumption/Marrakech/conprod10Y.htm, accessed 2 /8/07, copy on file

UNCED (1992) *Agenda 21: The United Nations Program of Action From Rio* (New York: U.N. Publications)

UNDP (1998) *Human Development Report* (Geneva: UNDP)

UNDP (2007) *Human Development Report 2007/2008: Fighting Climate Change: Human Solidarity in a Divided World* (Geneva: UNDP)

UNEP (United Nations Environment Programme) (2001) *Consumption Opportunities: Strategies for Change, A Report for Decision-makers* (Geneva: UNEP)

UNRISD (2000) *Visible Hands: Taking Responsibility for Social Development* (Geneva: UNRISD)

Vale, B. and Vale, R. (1977) *The Autonomous House: Design and Planning for Self-sufficiency* (London: Thames and Hudson)

Vale, B. and Vale, R. (1991) *Green Architecture: Design for a Sustainable Future* (London: Thames and Hudson)

Vale, B. and Vale, R. (2000) *The New Autonomous House: Design and Planning for Sustainability* (London: Thames and Hudson)

Van Sambeek, P. and Kampers, E. (2004) *NU-Spaarpas: The Sustainable Incentive Card* (Amsterdam: Stichting Points)

Van Vliet, B., Chappells, H. and Shove, E. (2005) *Infrastructures of Consumption: Environmental Innovation in the Utility Industries* (London: Earthscan)

Wackernagel, M. and Rees, W. (1996) *Our Ecological Footprint: Reducing Human Impact on the Earth* (Philadelphia: New Society Publishers)

Wakeman, T. (2005) 'East Anglia food link: An NGO working on sustainable food', paper presented at the *Grassroots Innovations for Sustainable Development* conference, UCL London, 10th June, 2005, http://www.uea.ac.uk/env/cserge/events/2005/grassroots/index.htm

Walker, P., Lewis, J., Lingayah, S. and Sommer, F. (2000) *Prove It! Measuring the Effect of Neighbourhood Renewal on Local People* (London: New Economics Foundation, Groundwork and Barclays PLC)

Walker, W. (2000) 'Entrapment in large technology systems: institutional commitment and power relations', *Research Policy*, 29(7–8): 833–846

Ward, B. and Lewis, J. (2002) *The Money Trail* (London: New Economics Foundation)

Waring, M. (1988) *If Women Counted: A New Feminist Economics* (San Francisco: Harper and Row)

WCED (1987) *Our Common Future* (Oxford: Oxford University Press)

Weber, M., Hoogma, R., Lane, B. and Schot, J. (1999) *Experimenting with Sustainable Transport Innovations: A Workbook for Strategic Niche Management* (Twente: University of Twente Press)

Wedge Card (2008) 'A Meeting With Mr Cameron', Weds 16 April, 2008 http://wedgecard.blogspot.com/, accessed 28/4/08

Whatmore, S. and Thorne, L. (1997) 'Nourishing networks: alternative geographies of food', in D. Goodman and M. Watts (eds) *Postindustrial Natures: Culture, Economy and Consumption of Food*, pp. 287–304 (London: Routledge)

White, N. (ed.) (2002) *Sustainable Housing Schemes in the UK* (Hockerton, Notts: Hockerton Housing Project)

Wilk, R. (2002) 'Consumption, human needs and global environmental change', *Global Environmental Change*, Vol 12(1), pp. 5–13

Williams, C. C. (2004) 'Compensating resident involvement: the "Just Rewards" campaign in the UK', *Planning, Practice and Research*, Vol 19(3), pp. 321–327

Williams, C. C. (2005) *A Commodified World? Mapping the Limits of Capitalism* (London: Zed Books)

Williams, C. C., Aldridge, T. and Tooke, J. (2003) 'Alternative exchange spaces', in Leyshon, A., Lee, R. and Williams, C. C. (eds) *Alternative Economic Spaces*, pp. 151–167 (London: Sage)

Williams, C. C., Aldridge, T., Tooke, J., Lee, R., Leyshon, A. and Thrift, N. (2001) *Bridges into Work: An Evaluation of Local Exchange Trading Schemes (LETS)* (Bristol: Policy Press)

Wilsdon, J. and Willis, R. (2004) *See-through Science: Why Public Engagement Needs to Move Upstream* (London: Demos)

Winner, L. (1979) 'The political philosophy of alternative technology', *Technology in Society*, 1, 75–86

Winter, M. (2003) 'Embeddedness, the new food economy and defensive localism', *Journal of Rural Studies*, 19(1), 23–32

Wrench, T. (1991) *Building a Low-Impact Roundhouse* (East Meon, Hampshire: Permanent Publications)

Wrench, T. (2001) *Building A Low-Impact Roundhouse* (East Meon: Permanent Publications)

WWF (2006) *Living Planet Report 2006* (Gland, Switzerland: WWF)

Yearley, S. (1988) *Science, Technology and Social Change* (London: Unwin Hyman)

Young, S. (1997) 'Community-based partnerships and sustainable development', in S. Baker, M. Kousis, D. Richardson and S. Young (eds) *The Politics of Sustainable Development*, pp. 217–236 (Manchester: Manchester University Press)

Zadek, S. and Evans, R. (1993) *Auditing the Market: A Practical Approach to Social Auditing* (London: New Economics Foundation and Gateshead: Traidcraft Exchange)

Zadek, S., Lingayah, S. and Murphy, S. (1998) *Purchasing Power: Civil Action for Sustainable Consumption* (London: New Economics Foundation)

Zadek, S., Pruzan, P. and Evans, R. (1997) *Building Corporate Accountability: Emerging Practices in Social and Ethical Accounting, Auditing and Reporting* (London: Earthscan)

Zelizer, V. A. (1994) *The Social Meaning of Money* (New York: Basic Books)

Index